Mountainous West, Denali to Pico de Orizaba

Roland H. Wauer

To order additional copies of this book, contact:
Xlibris
844-714-8691
www.Xlibris.com
Orders@Xlibris.com

ISBN: Softcover 978-1-6641-7967-7
Hardcover 978-1-6641-8145-8
EBook 978-1-6641-7968-4

Library of Congress Control Number: 2021912900

Print information available on the last page.

Rev. date: 07/15/2021

Contents

Dedication

To all those who love the mountains.

Introduction

> Climb the mountains and get their good tidings. Nature's peace will flow into you as sunshine flows into trees. The winds will blow their own freshness into you and the storms their energy while cares will drop off like autumn leaves. *–John Muir*

Mountains are far more than rocks, forests, and wildlife. They are a national treasure. They demonstrate interrelationships that are examples of all that is right in today's natural environment. There can be found wildness that provides for the falcon, deer, and bear. There is where those of us who love nature, can discover our own peace and understanding.

Stewart L. Udall wrote that "natural treasures are in reality a heritage of all mankind. They transcend provincial boundaries. They are a gift to those who prize the natural world and its healing influence."

Having lived in the western mountains much of my life, I have sampled many marvelous places from Alaska to Central America. Memories include such unexpected experiences as encountering a lynx in my path at Denali, being treed by a moose at Leigh Lake in the Grand Tetons, and admiring a peregrine at its eyrie on Loomis Peak in Mexico's Maderas del Carmen.

Travels to all of our national parks, places that are considered the very best of all our natural wonders, have honed my perspective of favored landscapes. Nowhere else claims first place than the Grand Tetons of Wyoming. Hikes to Cascade Falls and Lake Solitude, wanderings along the shore of Jenny Lake, and the searching for short-eared owls on Antelope Flat are but a few.

Grand Tetons & Snake River

The grandeur of North Cascade National Park provided yet another memory that is impossible to duplicate. Where else can one find black swifts speeding by in what seemed like utter uncontrol. How often were the descending and decelerating songs of canyon wrens echoing from surrounding cliffs? And watching an American dipper flying underwater in its search for insects was particularly special as well.

There also was the time on Angel's Landing in Zion National Park when I came face to beak with a spotted owl. And further on I watched a peregrine capturing white-throated swifts in mid-air; its timing was unbelievable.

The mountains provided many more memories that live in me. The opportunities to climb upward into the clean, fresh highlands have added immeasurably to my pleasures.

Acknowledgements

This book reflects my longtime love for the National Park Service. I worked for the NPS for 32 years in such remarkable areas as Crater Lake, Pinnacles, Death Valley, Zion, Big Bend, Great Smoky Mountains, and the Virgin Islands; in that order. A truly diversity of places and resources. Also, during those years, I was able to visit many additional natural areas.

Growing up in Idaho Falls, Idaho, it was the nearby Grand Tetons where I first was attracted to the mountains. Some of my most cherished memories were hiking the Teton trails. And I thank those park interpreters and rangers for their early introductions to beauty and inspiration.

This book would not have been written, however, if it were not for many friends and colleagues who provided their support during that early period and ever since. Although the majority of the photographs in Mountainous West are those I have taken over the years, I am especially grateful for those provided by my wife, Betty Wauer, my brother Brent Wauer, and Greg Lasley, who supplied several of the wildlife shots. All of those contributed photos are so indicated in the captions.

There are yet a few additional friends who gave their time and energy. I thank LeeAnn and Bill Nichols for their timely assistance in computer-support; and I also thank those who gave moral support to my recent activities, namely Bill, LeeAnn, and Barry and Sharon Nichols.

Chapter 1

Denali to the Rocky Mountains

From my perch at Camp Denali, I looked out at the huge sprawling base of America's most glorious mountain, Denali. At 20,320 feet elevation, it is the highest in North America. This morning, the ever-present cloud cover had not yet formed; I could see all of Denali. No other mountain claims such an enormous space. I could hardly believe I was so fortunate to view Denali on such a sparkling-clear day.

Today at Camp Denali was part of my second trip to Denali National Park (earlier known as Mount McKinley National Park). As a board-member of the National Park Conservation Association, we had met in Anchorage and travelled by train to park headquarters, from where we were bussed the 92-mile-long unpaved Wonder Lake roadway to Camp Denali, owned by one of our members.

Denali from Denali Lodge

Although marveling at my location so close to the great mountain, three unexpected things happened to me at Camp Denali: finding a lynx at my doorstep and an unexpected bird and butterfly on a short hike. I wrote about my lynx encounter in *When I Was Younger*:

One morning as I was leaving my cabin, heading to the lodge for breakfast, I discovered a lynx in my way. It was sitting just outside my door, gazing at me as I stared back at that magnificent, beautiful creature. Then I slowly eased back to where I had a camera and then eased back to the screen door. The lynx was still there sitting in the walkway as if it was waiting for me. I was able to take a single photograph through the screen, but when I tried to slowly open the door for a better photo, that was a mistake; it then walked away into the forest. It was my first and only lynx sighting and it was a thrill I will long remember.

Lynx through screen door

Although the majority of the three days at Camp Denali was taken up by meetings, the second morning was free to hike, photograph, or engage in whatever met our fancy. I chose to hike up a trail just behind my cabin. It was rather steep, but it was a good opportunity to stretch my legs and enjoy being alone. And as a long-time birder and a novice butterflier, I recorded all that I saw in my field-journal.

My first surprise that morning was finding a Say's phoebe. I was very familiar with this flycatcher from the southwestern deserts, such as Death Valley and the Texas Big Bend country, but I did not expect it in the highlands of Alaska. It sailed from perch to perch in a graceful manner, and each time it landed it gave its rather sorrowful two-part call. And as I continued up the trail, I encountered another surprise, a Milbert's Tortoiseshell, a butterfly that I later learned was a new record for Denali National Park. Truly a gorgeous creature!

Three such unexpected wildlife observations are what makes for memories that highlight any outdoor adventure.

An earlier visit to Denali occurred when Betty and I drove the 1,700-mile-long Alaska Highway, from British Columbia through the Yukon, to Denali National Park. Although the entire route offered marvelous scenery and extraordinary wildlife, the roadway itself, although "paved," provided us with a number of problems. Our initial plan was to carry most of the food needed for a four-week trip in our trailer, but after two broken springs, each requiring two- or three-day stops to send for replacements and repair, and a flat tire, we were forced to make adjustments. We ate the heavy canned goods first and bought only light-weight groceries thereafter. July on the Alaska Highway is when the roads are under repair following the long, hard winter months. I still have a tiny, amazingly sharp rock that pierced my tire.

Milbert's Tortoiseshell

In a sense, the Alaska Highway is a wildlife corridor. Elk were particularly common along the roadsides, moose were seen in many of the wetlands, and Dall sheep were reasonably common in places where the highways cut through high roadsides that provided them easy access and retreat. It appeared that the reason they spent time on the roadside was due to salt that had been used to help clear icy places in winter. Salt apparently was important to their nutrition.

The Dall sheep seemed unafraid when I remained in the truck, but they immediately retreated up the slope each time I left the truck to take a picture. Dall sheep are similar to bighorn which I am so familiar with from my years in Death Valley. But, of course, they possess white pelage, instead of grizzled gray-brown. The male's large, curved horns are similar. And body-sizes are about the same; bighorn's body size is, perhaps, slightly smaller. They nevertheless are equally impressive!

Our route to Denali took us from Bellingham, Washington, to Hope, Alberta, where we connected to Canada Highway 5 to Kamloops and on to Cache Creek where we met Highway 97 that took us all the way to Williams Lake and Prince George. Dawson's Creek, about 250 miles beyond Prince George, was the next town of any size. It was only there where I begin to feel that we truly were on our way to Alaska. Highway 97, north from Dawson's Creek, is known as the Alaska Highway. At Watson Lake, we turned northwest on Highway 1 that connects to Highway 2 at Tok Junction and southwest to Anchorage. Highways 1 and 2, especially between Haine's Junction and Tok Junction run along some of the most outstanding scenery I had ever seen.

Both Kluane National Park and Wrangell-St. Elias National Park offer high snowy mountains which led to numerous highway stops to gawk and take pictures. The tallest of the abundant peaks is Kluane's Mt. Logan, at 19,850 feet elevation. We stayed overnight at Haines Junction to restock and to absorb some of the character of the area. It was there where Betty purchased some local apples and baked an apple pie. Eating an apple pie while admiring the amazing scenery all around us was one of my most unexpected human experiences on the trip.

We spent another night at Kluane's Kathleen Campground. It was fairly busy as it was one of the few real campgrounds along that portion of the highway. However, travelers are allowed to stop at any of the numerous pull-offs along the highway. We took advantage of those opportunities for lunches and for overnights on several occasions.

Klaune NP, Yukon

View from Kathleen Campground

Gray Jay, by Greg Lasley

It was in Kathleen Campground where we became personally acquainted with gray jays. Although this personable bird had been seen on numerous occasions along our route, at Kathleen Campground it became almost a pest. It seemed to demand a handout. When I held a piece of bread above my head, one of the jays readily took it and flew off to some secret caching place.

Denali NP and roadway

In my later book, *Wild Critters I Have known*, I actually listed the gray jay as the "official avian ambassador" for Wrangell-St. Elias. Other avian ambassadors for Alaska parks included: willow ptarmigan for Denali, Kittlitz's murrelet for Glacier Bay, gray-cheeked thrush for Kobuk Valley, and golden eagle for Lake Clark.

Wrangell-St. Elias National Park lies just north of Kluane. Wrangell contains much of the same superb scenery. And it is the largest national park in all of Alaska, at 13 million acres. That is the same size as Yellowstone National Park, Yosemite National Park, and the country of Switzerland combined. Plus, Wrangell contains an enormous ice sheet, the Kennecott Glacier that lies on the slope of Mount Buckhorn (16,380 feet), the fifth tallest peak in Alaska.

Once we arrived in Denali, we were able to park our trailer in the small Wonder Lake Campground. This location inside the park allowed us to drive the roadway as we wanted; the majority of park visitors, in order to see the heartlands, must take a transit bus from the park entrance to Wonder Lake. Our opportunity to stay within the park was largely due to an earlier arrangement.

On one morning drive, while searching for birds and any other wildlife possible, I encountered a male caribou grazing very near the roadway. I was able to take a photograph from my vehicle, although when I did step out for another shot, it seemed to totally ignore me.

It was an impressing animal! The antlers were huge, suggesting maturity, and the pelage, at least from a distance, looked smooth and almost velvet-like. Its' shoulders and neck were a creamy white, and its' face and belly were a dark brown. European caribou are often domesticated and are then known as reindeer.

Caribou at Denali NP

From our campsite near Wonder Lake, I was free to explore the area alone. One early morning, although we were so far north that it was still twilight at midnight, I left the trailer and walked along the road looking for whatever birds that I might find. I had stopped at a groove of aspen to see what birds might be present, when suddenly, only about 50 feet away, was a wolf. It was just standing there, staring at me. It was a marvelous moment and one I will never forget. It did not appear afraid, but it seemed almost like it was trying to stare me down. Through binoculars I could actually see the bluish color of its eyes. I later told Betty that I had a feeling it was staring into my soul.

I was truly enamored with all of Denali. But maybe, except for the mountain itself, I was most impressed with the immensity of the scene. Everything was enormous. The open valleys, the low hills, and great distances in all directions.

Alaska's Lake Clark National Park is so isolated that it can be visited only by air. About 100 miles southeast of Anchorage, it is part of the Chigmit Mountains; the highest peak is Mount Redoubt at 10,196 feet. During the trip in which Betty and I drove the Alaska Highway, we stayed a few days in Anchorage. I was able to visit the Alaska Region Office of the National Park Service. And during that visit I was able to arrange a flight to Lake Clark. The arrangement was designed to leave me off at the Lake Clark office where I could spend two nights, while the NPS plane went on to visit other locations. I jumped at the opportunity.

Lake Clark, aerial

My two days at Lake Clark were marvelous. After leaving Betty in our trailer in Anchorage, I boarded a NPS pontoon plane and was soon en route to Lake Clark. The scenery was unbelievable! Along the way, the pilot showed me several key Alaska sites, such as Kenai Fjords National Park, and we flew over so many glaciers that it was difficult to remember them all. We landed on Lake Clark and taxied up to a small cabin with a small dock where I unloaded. The local park ranger was absent for a few days, so I was allowed to use the cabin. And he had left two still live salmon tied to the dock for my use.

I fried those two salmon with onions and salt and pepper that evening and ate more the next morning and for lunch and dinner and again the following morning before my ride appeared to take me back to Anchorage. Fresh salmon is my principal memory of Lake Clark that I have carried with me ever since.

The birdlife at Lake Clark was about what I had expected: Clark's nutcracker, gray jay, ruby-crowned and golden-crowned kinglets, and pine siskins. I also had an adult bald eagle flying along the shoreline. It came close enough that I was able to take a number of photographs.

Bald Eagle in flight, by Greg Lasley

I had an additional flight into the Alaska heartlands during the years that I was working in DC as Chief of Natural Resources for the National Park Service after President Jimmie Carter signed the Alaska National Interest Lands Conservation Act. It added 44 million acres in thirteen Alaska units of the National Park Service. It also added twenty-five free-flowing rivers to the National Wild and Scenic River System, and the nation's wilderness system increased by 56.4 million acres. Director Russ Dickinson asked me to go to Alaska and visit all the new areas so that there would be "someone in the Washington Office with basic knowledge of the new areas." I was of course elated!

My first stop was Anchorage where I visited with Regional Director John Cook about my trip. Then I was on to Fairbanks where I briefly visited with Fred Dean at the University of Alaska. Fred was involved with a number of projects throughout Alaska. The next morning, I boarded a Grumman Goose 789 aircraft, piloted by John Warner, to continue northward. I described that trip in *My Wild Life, A Memoir of Adventures within America's National Parks*:

> We flew over thousands of acres of tundra, mountains, rivers, and lakes first to Bettles, north of the Arctic Circle, then on to the Noatak; Gates of the Arctic; Kobuk Valley's dunes and Salmon River drainage; again, over Noatak's Grand Canyon; and eventually to Cape Krusenstern. We saw lots of country but got little more than a poor understanding of the area's true significance. I did attain a magnificent perspective of its grandness, vastness, and remoteness, but I had only minimal time on the ground. We did stop to visit Embryo Lake in the upper Noatak, Kurupa Lake in the northern edge of the Gates, and Tulugak Lake in the lower-middle Noatak Valley.

Alaska Range, aerial

At each of the three overnight stops, we pitched our tents, spent numerous hours wandering over the tundra and climbing onto the adjacent hilltops, and discussing various natural resource issues. The most pertinent topics included the Northwest caribou herd, Alaska's fire-management program, and the lack of research personnel and insufficient funds for Alaska, especially for the new park units. At each lake, Bill Palleck, park ranger from the Alaska Regional Office, spent the first hour or so fishing for our dinners. Bill's daily catch was another of my true highlights of the trip!

I spent as much free time as possible wandering about looking for birds. Some of the most exciting birds, for me, included a couple of Pacific loons; harlequin ducks, including a female with thirteen young; and a rock ptarmigan, hoary redpolls, tree sparrows, Lapland longspurs, wheatears, and snow buntings.

After a day in Kotsebue and a visit to Cape Krusenstern, we flew south to Deering and over Chimisso Wilderness, where I spotted a white morph gyrfalcon perched on a high rocky outcropping. I immediately asked John to circle back to where I could get a better look. It was still there; I was much impressed by its stature. Its white head and tail were obvious, and its snow-white back was marked with numerous black streaks.

Gyrfalcons are a bird of the far north with a range that includes all of Alaska and east to Greenland. It is a heavy-bodied bird, closely related to peregrines, but larger and more powerful. It very rarely occurs in the lower states, even during invasion years when its principal food supply in the north is in short supply.

That sighting of a gyrfalcon on that last day over northern Alaska was a fitting end to a wonderful trip. A memory I will cherish forever!

Chapter 2

Rocky Mountains

South of the remarkable complexity of Alaska's highlands is the northern end of the Rocky Mountains. Although impressive, the Rocky Mountains can hardly be compared to that which dominates so much of Alaska. Yet, six of the Canadian/U.S. national parks possess truly gorgeous landscapes: Banff, Jasper, Kootenay, Yoho, Waterton Lakes, and Glacier. Glacier and Waterton Lakes are members of the Glacier-Waterton International Peace Park. I described the region in *The Visitor's Guide to the Birds of the Rocky Mountain National Parks*, thusly:

> Banff is an area of mountains, valleys, glaciers, forests, alpine meadows, lakes and wild rivers along the Alberta flank of the Continental Divide. The highest elevation is 11,500 feet, and much of the area is above treeline... Jasper also lies along the eastern slope. It is one of Canada's largest parks at 2,688,000 acres...an area of broad valleys and rugged mountains with a parkway that is unparalleled for beauty as it runs alongside a chain of massive icefields straddling the Continental Divide. The park's maximum elevation is 12,294 feet at the summit of Mount Columbia, which is the highest point in Alberta. Jasper's lowest river valleys are approximately 3,000 feet in elevation, producing the greatest relief in any of Canada's Rocky Mountain parks.
>
> Kootenay is described as "a land of startling contrasts of towering summits and hanging glaciers, narrow chasms and color-splashed mineral pools... Yoho is a park with snow-capped mountain peaks (more than 25 are over 10,000 feet), roaring rivers and deep silent forests."

Banff NP, Alberta

My favorite of the five Canadian parks is Banff. As a kid with my parents, I visited Banff several times; I considered Lake Louise one of the world's most gorgeous settings. My time in Banff also produced some favorite wildlife memories: elk, moose, deer, and black bear were commonplace, and I once encountered a willow ptarmigan that allowed a surprisingly close-up inspection. I later described that bird as follows:

> Willow ptarmigans possess an all-black tail; even in mid-winter, when their plumage turns all-white to help hide them from their enemies, their tails, eyes, and bills remain black. Summer plumaged birds are brown with a reddish tint and have white wings. During courtship, the male's bright red comb, the bare skin above the eyes, swells to four times the normal size.

Golden Eagle in flight

I also recall an interesting morning on Waterton's prairie grasslands, just above the Red Rock Canyon Road junction. At least a hundred Columbian ground-squirrels "inked" at me. And they got especially excited when a golden eagle cruised by only 20 to 30 feet above the ground.

Columbian ground squirrels are fairly large ground squirrels with a reddish-brown nose and forelegs. And their bushy tail also is reddish and edged with white. They typically live in colonies in alpine meadows and grassy lowlands.

Glacier National Park, also part of the Canadian Rockies, straddles both Alberta and Montana. These dual areas contain a crossroads of a wide variety of floristic regions that result in a higher diversity of plant and animal species than occurs anywhere else in the northern Rocky Mountains. Glacier's high point is Mount Cleveland at 10,466 feet. And Going to The Sun Highway, which runs for 50 miles between the St. Mary entrance and West Glacier, provides some awesome scenery. St. Mary's Lake is one of the best known of Glacier's abundant landmarks.

Our drive into Jasper, along the Icefield Parkway, was spectacular. And after finding a campsite, we spent the remainder of the day visiting Maligne Lake and Cottonwood Slough. At Cottonwood Slough, I was amazed to find five species of Empidonax flycatchers (alder, willow, least, Hammond's, and dusky). And the yellow-rumped warblers at Cottonwood Slough included both the western Audubon's and the eastern myrtle forms. I had finally discovered a place here the two forms overlap, where they interbreed and where ornithologists gained sufficient insight to lump these two birds into a single species.

Going to the Sun Highway

St. Mary's Lake, Glacier NP

On another day, we took the tram to the top of The Whistlers, an area above treeline that serves as a well-known wintertime ski area. Although I did not find an abundance of birds at this open highland area, a pair of common ravens put on a wonderful example of aerial maneuvers for us. They flew side-by-side for a while and then soared upward for a couple hundred or more feet before diving downward and rolling in unison before turning upward once again. They put on a truly marvelous show!

Pacific Coast influences are most obvious along the lower, western slopes, where various prairie grasses, golden aster, moss phlox, early cinquefoil, and prairie rose are common. Mountain tundra and rocky habitats dominate the highlands…forest occurs below approximately 7,500 feet.

I was excited to find gray-crowned rosy finches on the tundra and along the edges of snow fields and rocky ledges. They can be incredible tame; I was able to walk right up to a feeding bird. I later described them in *Visitor's Guide*, thusly:

> Although they are about the same size as a white-crowned sparrow, their appearance and behavior are very different. Rosy finches possess cinnamon-brown plumage tipped with a rose-color on their rumps, forehead, shoulders, and underparts, and a gray crown with a black forehead. In sunlight, at the edge of a glistening snow patch, they possess a subtle but marvelous beauty.

Glacier NP, Alberta

One day I hiked to Avalanche Lake, 2.4 miles above the Avalanche Campground, where Betty and I were camped. It had rained the night before and the trail was wet; the streamside vegetation was still dripping with rainwater. But the sky was crystal clear and bright blue. Patches of snow still clung to crevices and ridges on the cliffs that rose several thousand feet above the narrow valley in which the trail lies. I wrote about that day as well:

> From almost anywhere along the trail the eerie, bell-like songs of varied thrushes were heard. Most of the songs rang from the foliage of western redcedars, and most of the birds were hidden from view. But on two occasions a brightly colored male sang from an open perch near the treetops. These glamorous robin-sized birds looked all the word like bright ornaments atop a green Christmas tree. The varied thrush sports a bright orange breast with a coal black chest band, orange eyebrows over bold black cheeks, and gray-blue back and wings that possess two orange wing bars.

On another day, I hiked up to Avalanche Lake to again enjoy the beauty of the scenery. I sat at the edge of the lake admiring the crystal-clear water and the amphitheater-like surroundings, a matrix of black and gray cliffs, green vegetation, and blue sky. Then suddenly, less than 50 feet away was an American dipper floating on the lake's surface, consuming a rather large insect of some sort. Then, just as suddenly, it dived beneath the surface. When it bobbed up a few seconds later, it had captured another larva. When it dived again, I stood up so that I was able to watch it swim below the surface of the lake. It actually propelled itself through the water with its wings; its feet hung unused. I watched it grab another larva and swim to the surface. It immediately consumed its catch. But this time it paddled to a nearby log protruding above the surface, walked a few inches up the log, and proceeded to preen. It was a remarkable bird, very plump and short-tailed, with all gray-brown plumage.

Mount Revelstoke and Glacier national parks occur in the Selkirt Range, just west of the Rockies, part of the Columbia Mountains of British Columbia. This region of North America is often referred to as the "interior wet belt." Average annual precipitation at Revelstoke is about 28 inches compared with only 12 inches at Calgary, Alberta, on the eastern flank of the Rocky Mountains. They are, however, an interface with the more southerly Rocky Mountain parks.

Mount Revelstoke National Park is best known for its topographic relief that runs from 1,500 feet at the town of Revelstoke, on the Columbia River, to 8,681 feet at the summit of Mount Coursier, providing a total relief of 7,180 feet. The park includes more than 50 miles of trails, including the spectacular but short Mountain Meadows Interpretive Trail at the summit. And I also found Revelstoke's Skunk Cabbage Nature Trail very productive. It was there where I actually found four pairs of singing willow flycatchers. I was surprised at their abundance within such a fairly small wetland.

And at nearby Maskinonge Lake, I spent time watching black terns. The behavior of the 10 to 12 individuals was extremely acrobatic. One individual, carrying a small fish in its bill would hover in flight and then dive, very gracefully like a swallow, and then passed its catch to another, assumedly an act of courtship. Their all-black bills and plumage, except for their whitish upperwing coverts and wing linings gave them a stately appearance.

Hidden Lake Trail, Glacier NP

At Glacier, we walked the Hidden Lake Trail, which starts just behind the Logan Pass Visitor Center, and passes through an open field of wildflowers. Millions of alpine flowers dotted the landscape. The bright reds, purples, and yellows were in sharp contrast to the velvet-green sedges, lichens, mosses, and algae growing at ground level.

I was surprised on finding a white-tailed ptarmigan along the trail, where dozens of visitors passed by daily. I didn't see it until I was only a few feet away because it blended so well into the ground cover. Its mottled plumage of blacks, browns, golds, and whites provided it with amazing camouflage. Through binoculars I could see the scarlet eye combs above each eye. Its snow-white belly and wings were partially hidden from view by its crouched position.

And when walking Glacier's Loop Creek Trail, my highlight bird was the black swift. We gazed in admiration as 15 to 20 individuals performed their amazing acrobatics directly above us. For two to three minutes they were in sight before moving on to other locations. It was an exciting encounter!

Logan Pass, Glacier NP

Subalpine forest dominated the upper slopes, forming extremely dark green forests above the lush valleys, between avalanche paths and below the high gray cliffs. Common birdlife in these areas included the rufus hummingbird; willow flycatcher; yellow, MacGilivray's, and Wilson's warblers; and song and fox sparrows. I wrote about fox sparrows in *Visitor's Guide*:

> The fox sparrow is the largest of North America's 38 sparrows and is blessed with one of the most beautiful but complicated of songs. John Terres calls it "clear, exultant, melodious, flutelike." It usually begins with an introductory whistle followed by a series of sliding notes, whistles, and slurs. They possess three or more song types, and if uninterrupted will sing all types one after the other until their entire repertoire is completed; it will then start over.

One June morning I described a series of a fox sparrow's songs in my field notes: "tee-d-dee-dee-dee, sweet sweet-tree." With emphasis on the "tree': then, "swee-dee-deet, che-che-che"; "sweede, dee-de-de-de-de, chip"; and finally, "tur-de-wee-dee. Tzee. Chip, de."

Fox sparrows are best identified by their large size, uniform dusky-brown plumage, and pale, heavily streaked underparts.

Old Faithfull, Yellowstone NP

Yellowstone Falls, Yellowstone NP, by Betty Wauer

Yellowstone and Grand Teton national parks also are part of the Rocky Mountains. The majority of Yellowstone is situated on a high central plateau between the Absaroka Range on the east and Gallatin Range on the northwest. Most of the park lies between 7,000- and 9,000-feet elevation, but Eagle Peak, the highest point in the park, reaches 11,358 feet above sea-level.

Yellowstone is one of our most popular parks, and to most visitors, Old Faithful Geyser is a required visit. That site can be extremely busy and seems to satisfy most folks; once they see Old Faithful, they've seen Yellowstone. And Yellowstone Falls is often only a secondary interest. For me, Yellowstone Falls is equally superlative.

Yellowstone also is an amazing wildlife park, one of the few places where one possibly can see both grizzly bears and wolves. I recall one visit when Betty cam-recorded a grizzly bear running across a distant meadow at a speed that I could hardly believe. It reminded me of a galloping horse. I've been told that it is impossible to outrun a grizzly; I believe it!

Yellowstone also is one of the few areas where wild bison can still be found. They roam the grassy hills and often frequent the roadsides where, because so many visitors stop to watch and photograph them, they often create a bison-jam. Park rangers have an almost impossible task, trying to move them off the highway and keep people from getting too close for a photo. There actually are cases when parents have tried to put their child on a bison for a photograph; far more visitors are injured from bison than by bears.

Bison in Yellowstone, by Betty Wauer

On one visit to Yellowstone, Betty and I camped at Fishing Bridge. It was there where I had a close-up encounter with fishing white pelicans. They did not dive on prey like brown pelicans, but fish from the surface by lowering their long bills and heads below the surface and literally scooping up their prey. I watched five individuals herding fish ahead of them into shallow water, where they were better able to scoop them up into their gular sacs.

Having grown up in nearby Idaho Falls, Idaho, I have visited the greater Yellowstone area on numerous occasions. Although it was never my favorite national park – the Tetons and Grand Canyon were favored – I did recognize early-on the important of the area as a truly significant wildlife preserve. And years later, while working as Chief of Natural Resources for the National Park Service in Washington, D.C., I again became involved in Yellowstone on a number of occasions. As chairman of the Grizzly Bear Steering Committee, I was intimately involved with maintaining endangered status for grizzly bears.

I learned that grizzly bear numbers had declined in the past 280 years from 50,000 occupying the entire area west of the Mississippi River to fewer than 1,000 bears occurring in only six isolated localities. Of these six places, only the Greater Yellowstone and Northern Continental Divide ecosystems contained populations large enough to exist in self-perpetuating conditions.

White Pelican

Grizzly Bear and Dick Knight

After a trip to Yellowstone to see for myself, and a long discussion with Dick Knight who was studying grizzly bears, I prepared a memorandum to members of the Steering Committee. I discussed my Yellowstone visit, including a Congressional Hearing I attended on grizzly bears, and I wrote the following:

> We no longer have the luxury of time to research the remaining parts of the puzzle. The Yellowstone grizzly bear picture is presently clear enough! Any additional delays in mitigations, actions that a bear management community has known it must take for several years, will likely result in the loss of grizzly bears in the greater Yellowstone ecosystem. It is imperative that highest priority be given to eliminating grizzly bear mortality. Only immediate and broad scale protection can save the grizzly. Increased protection efforts must be united by all the pertinent land managers. Without it we will only be documenting the demise of the grizzly bear within the Yellowstone ecosystem.

My memorandum was sent to all members of the steering committee as well as several conservation organizations. Within two weeks my memorandum appeared in numerous newspapers, from the East Coast to the West Coast. And the Park Service and various member were besieged with letters and editorials expressing concern. The New York Times, for instance, reported that the Yellowstone grizzly was "imperiled."

My position as chairman of the Steering Committee was also imperiled. I had given too much visibility to this issue; Secretary of Interior James Watt soon appointed a new chairman.

Bull Elk

I also visited Yellowstone after the 1988 fires which burned approximately 194,000 acres (36%) of the park. Much of the forest was affected, and yet I was able see much more of the landscape; driving many of the highways prior to the fire was like driving through a tunnel of lodgepole pines. Another memory from my drive through the burn that day was watching elk eat pieces of charcoal. It apparently provided them with essential nutrients. Their loud crunching was evident from a consider distance.

Elk are primarily grazers, like cattle, but they also browse on leaves of shrubs and trees. Aspens are favored when available; they utilize both the leaves and new stems. Elk eat constantly and consume 8.8 to 14.4 pounds of vegetation daily. Wintertime is hardest on them when they must clear the snow by pawing to reach grasses beneath.

There are a number of localities where elk gather during the winter months. One of those is Jackson Hole, Wyoming, in a huge meadow area just north of the town of Jackson. As a youth, my family occasionally, before heavy snow limited our trips, drove the approximately 100 miles to see the enormous herds of elk. Those herds could exceed 100,000 individuals, about one-half of the total elk population in the Greater Yellowstone/Teton ecosystem. Elk from as far away as Yellowstone National Park may spent as much as six months there in the Jackson Hole Elk Refuge. The U.S. Fish and Wildlife Service feed the elk each winter. I wrote about the elk refuge in *Wild Critters I Have Known*, thusly:

> During the time that the large herds of elk are utilizing the Jackson Hole Refuge, they shed their antlers. And local Boy Scouts take the opportunity to collect the discarded antlers which they auction off to interested buyers. In 2010, they auctioned off 5,600 pounds of antlers, collecting $46,000 of which 80% was retained by the refuge for management and feeding the elk.

It takes a good deal of grasses and herbs to sustain an elk, considered one of the largest land animals in North America. An elk can weigh 325 to 1,100 pounds, and measure four to five feet at the shoulder. They are one of the ruminant species, with a four chambered stomach, meaning that, like cattle, they lie down to process the foods.

Grand Teton from Antelope Flat

My favorite of all the Rocky Mountain parks is Grand Teton. During my youth, my family spent a couple weeks each summer in the Tetons. I often hiked across Jenny Lake to Cascade Canyon, and few times beyond to Lake Solitude. On one occasion, I climbed onto Grand Teton (see cover), the park's high point at 13,770 feet elevation; I reached the Teton Glacier, located just below the summit.

Much of the Teton Range consists of bare rock; treeline is at about 10,000 feet elevation. Three broad vegetative zones occur within the park: alpine, subalpine, and montane, each containing a variety of plant and animal communities. Alpine grasses, herbs, and sedges form low-growing tundra-like vegetation on soils above treeline. Dwarfed subalpine fir and Engelmann spruce occur in scattered patches above the taller more extensive spruce-fir forests that range down to about 6,500 feet elevation. Douglas-fir are scattered on lower ridgetops, and lodgepole pines dominate the glacial moraines at the base of the mountains.

Jenny Lake Trail, Grand Teton NP

The loveliest sounds of the forests, from my perspective, are two thrushes: Swainson's and hermit. These two birds occur throughout the spruce-fir communities where their flute-like songs are commonplace, even though the listener may not know where those wonderful renditions originate. I included comments about these thrushes in *The Visitor's Guide to the Birds of the Rocky Mountain National Parks*:

> Swainson's thrushes sing an ascending spiral of mellow whistles, while hermit thrushes sing a series of warbling notes that are repeated at different pitches. Both can sing for long periods of time, especially during the morning and evening hours. They are both slightly smaller than their cousin, the American robin, and possess a brownish back with whitish underparts with darker spots or streaks. Hermit thrushes possess reddish tails and whitish eye rings; Swainson's thrushes are slightly larger and have all-brown backs, buff eye rings and lores. They otherwise are difficult to tell apart, especially in the shadowy forest in which they reside.

Another of my favorite birds in the Tetons, and one that I admired so often at Jenny, String, and Leigh lakes, was the osprey or fish hawk. I wrote about watching a fishing osprey at Leigh Lake in *Bighorn, Pelicans and Wild Turkeys*, thusly:

> An osprey suddenly dropped from its hovering flight, diving almost straight down for about 100 feet, its great talons extended forward, its tail spread, and wings slightly folded. It hit the water with a huge splash and, for a second or two, totally disappeared underwater. Then it emerged with a great thrashing of wings and was soon air-borne once again. It tightly held a 10- to 12-inch trout by both talons, a moment later in midair, it shook vigorously to remove the remaining water from its black-and-white plumage. Then it rearranged its catch so that its head pointed forward to create less wind resistance. With only a slight labored flight, the osprey headed to its nest somewhere along the shoreline.

Jenny Lake, Grand Tetons NP, by Brent Wauer

Osprey, by Greg Lasley

As a youngster, my family always stayed at the Square-G Ranch (since removed) on the eastern side of the park. From there we were able visit the chain of lakes below Jackson Lake: Leigh, String, and Jenny lakes. On one hike to Leigh lake, I was treed by a cow moose, a harrowing experience at the time and always worthy of an exciting moose story. I included that encounter in my *Wild Critters I Have Known*:

> That incident began when our party of three hikers, including my dad, another boy about my age, and me, 12 or 13 at the time, came head to head with a moose on the narrow trail. I had been in the lead of our small group, and on seeing the bull moose in the trail, I raised my camera to get a photograph. I was too slow to get a quick shot of the bull (that quickly ran off the trail), but the cow moose was quickly filling the frame of my camera. Its image size was increasing dramatically. I instinctively knew that the smartest thing to do was to immediately leave the trail and move into the adjacent shrubbery.
>
> It was then that my dad yelled "quick, get up a tree." The three of us quickly grabbed a nearby tree and began to climb. Any way to escape the charging moose. We had no trouble finding a tree; the forest was full of lodgepole pines. Although lodgepole pines are rather thin and fairly easy to climb, the spiny trunks are not made for comfort. We nevertheless managed to shinny up eight to ten feet where we held on for dear life. The three trees were only ten to fifteen feet apart and barely ten feet off the trail. My camera was dropped below my tree.
>
> The moose charged at us, running unbelievably fast through the forest, weaving between the lodgepole pines like a snake through grass. It looked immense from my precarious position. It made more than a dozen charges, briefly stopping between each to snort and paw the ground, and her hackles were raised like that of an Andalusia bull. Yelling seemed to have no effect, although we discovered that sharp, loud whistles caused it to hesitate before making its next charge.

The charges continued for 12 to 15 minutes (it seemed like hours), but after the first six or seven charges it began to circle our trees, stopping along the edge of its route before charging back toward our trees. Her circle gradually became wider, and finally it reached the edge of Leigh Lake where it apparently had a calf hidden in the undergrowth. She then gathered up a calf and moved down the trail and out of sight. Once we saw the calf with its mother, we understood the mother's concern for its safety. That was the reason it had attacked us encroachers to its territory.

We waited several minutes before we decided to climb down from our trees onto the ground. It was then that we discovered that our arms were bloody, not from an injury from the moose, but from the tiny prickles of the trees. And we soon discovered that some of the blood was also from the untold number of mosquitoes that had taken advantage of our predicament. The mosquitoes were unbelievable, probably more than thousands. I remember rubbing my arms from the scratches received from the pine trunk bristles, and I then discovered my bloody hand and t-shirt. Even my open neck and face were covered with mosquitoes and bites. And my t-shirt was torn from the prickles on the tree trunk while holding on so tight.

Back on the ground, we began congratulating one another for still being alive. And we were about to hike back to our cabins at the Square-G Ranch, where we had been staying, when my dad suddenly said, "Listen, I hear it coming back." And he added, "Let's find better trees to climb." We then scampered about selecting more climbable trees and were soon perched a bit higher on our chosen trees than before.

We waited for several minutes, clinging on our trees, hardly breathing and listening in anticipating of another attack. But the moose did not appear. Finally, my dad called off the wait, admitting that he was wrong. He realized that the sound he heard was not that of the returning moose, but his own heartbeats that had sounded to him like the hoof beats of our antagonistic moose.

String Lake, Grand Teton NP

Following the Continental Divide southward from the Tetons, it passes through Wyoming and into Colorado and Rocky Mountain National Park. And beyond, in Colorado and New Mexico, are Mount Elbert, at 14,440-feet, the highest point in the Rocky Mountains, and the Sangre de Christo Mountains with 14,334-feet-high Blanco Peak that straddles the two states of Colorado and New Mexico.

Rocky Mountain National Park is special for several reasons. No other national park in the continental U.S. provides vehicular access to miles of tundra habitat. Spectacular Trail Ridge Road, which passes over "the roof of the Rockies," remains above treeline for eleven miles and reaches an elevation of 12,183 feet. The views are spectacular! To the south is the rugged Front Range while to the north, across the Fall River Valley, the view is dominated by the majestic Mummy Range. Far to the west is the Never Summer Range. The park's highest peak, to the south, is Long's Peak at 14,255 feet, but almost 100 peaks within the park's 265,668 acres exceed 11,000 feet elevation.

I recall one very cold morning on the tundra along Trail Ridge Road, searching for one of the park's most charismatic birds, the white-tailed ptarmigan. This is a hardy species that remains on its tundra homeland year-round. In the fall, it molts its mottled black-and-gold plumage to all-snow-white, except for its coal black bill and eyes and red eye combs. It also develops heavily feathered legs and feet that serve as snowshoes. It took me a good part of an hour, on that very cold, windy day to locate my bird. When I did, instead of flying or running away, it froze in place. It looked all the world like a golden ball of feathers, expect for its bulging bright red eye combs. Truly an amazing creature!

Trail Ridge Road, Rocky Mountain NP

White-tailed Ptarmigan

Another bird of interest that day on the tundra was the American pipit. I found several on that open, rocky terrain. Many displaying with their distinct behavior and territorial song, a long series of "treet" or "pip-it" notes. Some sang from rocky perches, but most would fly upward almost vertically to as much as 200 feet and float downward with their wings out, leg hanging, and tail upward, singing all the while. A remarkable courtship flight.

Pika

I also found two mammals in that tundra habitat, a yellowbelly marmot, a big fat yellowish-brown creature, and a pika. They seemed to have a very different personality and behavior. While the marmot sat on a bare rock at a considerable distance, it seemed too lazy to do much else. But I was mostly attracted to the pika by its sharp "heck" calls. It took several minutes to find it. It was a tiny little mammal, especially in comparison with the marmot, and its grayish-brown coat blended extremely well with the rocky landscape. Pikas are closely related to rabbits; they spend much of their summertime gathering grasses into little haystacks that they utilize for food during the winter months where they remain underground all winter.

I recall one "haystack" in the Beartooth Mountains of Montana that may have been nine inches tall, mostly of grasses, but it also contained an ample number of twigs, moss and lichen. It is estimated that the amount of plant material gathered each season would be enough to fill a bathtub. Pikas experience very short alpine summers, and most winters in the high mountains are extremely cold with snow covering much of their talus homelands.

Treeline, like that at Yellowstone and Grand Teton, occurs at about 11,000 feet elevation at Rocky Mountain. All the bare rock and broad expanses of tundra habitat above that fall within the alpine zone. This is the environment that Ann Zwinger and Beatrice Willard wrote so intimately about in their excellent book, *Land Above the Trees*. An interpretive sign along the Tundra Nature Trail features one of Zwinger's descriptive quotes: "This alpine tundra is a land of contrast and incredible intensity, where the sky is the size of forever and the flowers the size of a millisecond."

One evening at Rocky Mountain Park, I circled Bear Lake, following the self-guided nature trail along the rocky shore. And true to the park's brochure, the three common birds were Steller's and gray jays and Clark's nutcrackers. All three greeted me at the tailhead. The jays landed almost at my feet, expecting their usual handout of junk food that, in the long term, makes them more dependent upon human foods and less likely to survive harsh winters. Steller's jay were particularly colorful birds, with their dark blue bodies, blackish-blue heads, and tall crests. They were so close I could see the four white streaks on their foreheads.

Treeline, Rocky Mountains

Clark's Nutcracker, by Greg Lasley

I have had great affection for Clark's nutcrackers, ever since working at Crater Lake, where they are so common along the Rim. Not only do they frequent areas of high human use, but they also possess an exceptionally keen memory. Nutcrackers are able to remember the location of about 1,000 seed caches from one season to another. I wrote about this ability in *The Visitor's Guide to the Birds of the Rocky Mountain National Parks*:

> In fall, when conifer seeds ripen, nutcrackers pry the seeds from the cones in crowbar-fashion, with their sharp, heavy bills, then hide the seeds on south-facing slopes for winter use. They possess a special pouch under their tongue in which they are able to carry up to 95 seeds per trip. A study by ornithologist Stephen Vander Well, of Utah State University, proved that nutcrackers were able to recall where they had cached their seeds. The birds remember where the seeds are in relation to central landmarks, such as rocks. If the landmarks are moved, the areas the birds search are displaced an equivalent amount.

Great Sand Dunes

The Great Sand Dunes, located in south-central Colorado, are situated at the base of a great arc of the Sangre de Cristo Mountains. They rise to more than 14,000 feet elevation and serve as a magnificent backdrop to Great Sand Dunes National Monument. The dunes cover an area of almost 25,000 acres, and they are the tallest in North America, reaching almost 700 feet. Further south in the Sangre de Christos is Mt. Baldy that rises to 12,441 feet elevation.

During the years that I lived in Santa Fe, New Mexico, I hiked Mt. Baldy on numerous occasions. Located in the adjacent Santa Fe National Forest, Mt. Baldy, also known as Santa Fe Baldy, provides a significant watershed that empties into the upper Rio Grande. A portion of the region also is within the Pecos Wilderness.

I taught an ornithology class at the College of Santa Fe, and on one occasion I guided my class on a birding hike onto Mt. Baldy. It was a beautiful late spring day, and all the neotropical migrants had arrived on their nesting grounds. The most impressive bird that day was an olive-sided flycatcher that sang its clear "quick-three-beers" call over and over. My class was much impressed, and I recall that it became the class theme the remainder of our sessions.

Santa Fe Baldy

Chapter 3

The Cascades

North Cascades National Park sits at the far northern edge of the Cascade Range that runs south through the states of Washington, Oregon, and into northern California at Lassen National Park. Mount Adams, Baker, Hood, Jefferson, Shasta, and Lassen are all included in this majestic series of volcanic peaks.

North Cascades National Park encompasses an area of about 648,000 acres of which 93 percent is officially designated as the Stephen Mather Wilderness Area. The park brochure claims that the park "contains some of America's most breathtakingly beautiful scenery – high jagged peaks, ridges, slopes, and countless cascading waterfalls." It further states that the area "encompasses some 318 glaciers, more than half of all glaciers in the contiguous United States." I can't argue with those facts. And from my perspective, North Cascades, more than any of the other national parks, represents true wildness and amazing beauty.

North Cascades NP

Everywhere I went in North Cascades, black swifts seemed to be commonplace. The reason was that they nest behind the park's abundant waterfalls, where there are no competitors and predators. Nesting black swifts gather moss, algae, ferns, and other nesting materials in flight, which they glue together with their saliva; they use conifer needles and fine rootlets to line their nests. And they possess an unbelievable life history.

They are featured in the park's visitor center exhibit which includes a replicate of a nest with one tiny bird. The text with the exhibit reads: "Black Swifts fly up to 600 miles a day in search of insects to feed their one chick. While the parents are gone the chick waits, sometimes 2 or 3 days, in torpor, a state of prolonged sleep. Its breathing and heart rate drop dramatically to conserve energy, down to one breath and four to eight heart beats a minute."

Another northwestern specialty of the park is a little yellow-and-black bird with a rather distinct song. Townsend's warbler. Its song echoed from the foliage of a conifer some distance from where I was standing. It was a weezy but lovely song with three rising notes followed by two sharp descending ones, like "bzeee bzee bzee, tsee-see." Just as I zeroed in with my binoculars, it put its head back and sang again. Its coal-black throat, cheeks, and cap contrasted with the bold, canary yellow lines above and below the cheeks. Marvelous!

One day I walked the Cascade Pass Trail that switchbacks through montane forest for about two miles before coming out onto talus slopes with scattered subalpine firs and deciduous thickets. Forest birds detected there included Pacific-slope flycatchers, chestnut-backed chickadees, golden-crowned kinglets, hermit and varied thrushes, and yellow-rumped and Townsend's warblers.

Hoary marmots were reasonable common along the trail beyond the forest. One individual perched on a rocky outcrop and called sharp whistle-notes at me. Pikas called from the open scree slopes, and when a common raven soared over the pass and cruised across the open slopes, eight or nine pikas took up the call.

North Cascades NP

I sat at the pass for a long time admiring the incredible scenery in every direction. The Stehekin River drainage lay before me to the southeast, and the steep slopes, still partially covered with snow, provided a contrasting perspective. I then began to realize that I had been hearing high-pitched chipping sounds somewhere below me. With binoculars, I scanned the open slopes and low shrubbery. Nothing. Shifting to the right, I was suddenly staring at a little rosy bird picking at the open ground beneath a snowbank, 200 feet or more away. It was a gray-crowned rosy-finch, a true alpine species that spends its summers above treeline.

It gathers together in huge flocks after nesting, however, and moves to lower, warmer elevations to the east of the Cascades for the winter months. As I zeroed in on my little rosy-inch, with a black crown and light gray nape, I discovered six others, all feeding on the ground or on low shrubs. One individual seemed to be eating buds from a patch of red heather. Then, as I watched, they suddenly took off, flying toward me and up the steep slope to my right. In passing, I detected a series of high-pitched notes, "chee-chee-chi-chi-che."

Further south in Washington state is the grand jewel of the Cascades, Mount Rainier. Few mountains possess the grandeur and scenic splendor. It rises above the deep green forests like a great silver crown. Rainier, at 14,411 feet, is the highest and most impressive of the several majestic volcanic peaks in the Cascade Range. Peter Farb, in *Face of North America*, wrote:

> What makes Rainier so impressive is that it rises upon its immense base, covering about a hundred square miles, directly out of the tidewater of Puget Sound. On its flank, like ermine robes on the shoulders of some king of mountains, lie 26 glaciers, a greater expanse of ice than on any other peak in the mainland United States. Its majestic bulk, its isolation from the rivalry of other peaks, the fact that nowhere can the summit be seen from the base – these things have made Rainier the most striking of all the mountains in the mainland Unites States.

Mount Rainier

I have been extremely impressed with the area's natural features. Below its icy countenance, are lush coniferous forests, kept fresh by the Pacific influence, and rugged canyons and many streams that run off the highlands to join with rivers that run to the sea.

The forests vibrate with the sounds of birds – varied thrush, chestnut-backed chickadee, winter wren, and Townsend's and hermit warblers - that are abundant there. And a watchful nature lover who loves the abundant swift waterways may also see one of the bird world's most colorful ducks, the harlequin duck. I described this bird in *Birds of the Northwestern National Parks*, thusly:

> It sports an overall slate blue body with chestnut stripes on its head and sides, white markings scattered here and there, including a large crescent on its face, stripes and spots on the neck, and a collar. The name harlequin was derived from a comical Italian character of the theatre who wore mask and multicolored tights. Females are nondescript brownish birds with spots on their cheeks and head.

Rainier is the southern edge of this bird's range that extends northward on a broad band into northern Alaska. And it also occurs along the rocky coastline in northeast Canada and Greenland.

Crater Lake NP

Crater Lake lies to the south of Rainier, but lacks the height to be recognized for its elevation, although Mt. Thielsen, located just north of the park rises to 9,183 feet. Crater Lake is a product of a violent volcanic eruption about 7,700 years ago, when 12,000-foot-high Mount Mazama expelled great amounts of pumice and ash around the mountain and across vast stretches of North America. There is no evidence that the mountain exploded outwards, but with much of its innards gone, the mountain collapsed downward (or inward upon itself), leaving a great caldera or basin, with a maximum width of six miles. The caldera eventually filled with water from rain and snow to form Crater Lake.

Crater Lake is rather special to me because it was my first park when working for the National Park Service; I still have fond memories of the scenery and wildlife. Two birds are especially memorable: Clark's nutcracker and peregrine falcon. Clark's nutcrackers are unquestionably the park's most obvious wildlife. In *Birds of the Northwestern National Parks,* I wrote the following:

> These medium-sized, gray, black, and white birds, with conspicuous white trailing edges to their black wings, are widespread along the Rim, perched on various snags or on the low stone walls along the parking area and sidewalks, or flying by. If they are not immediately apparent, their guttural and drawn-out "kr-a-a" or "chaar" calls are commonplace. It is doubtful that any visitor to this marvelous park of the extraordinary deep blue lake will miss Clark's Nutcrackers.

Crater Lake, Rim view, by Betty Wauer

Peregrines seemed to spend an inordinate amount of time soaring along the Rim. Usually, these marvelous birds attracted attention by their high piercing calls. And often one or a pair of peregrines was visible relatively close by, either flying along the Rim to some peregrine destination or actually playing directly overhead. I wrote about a pair of Crater Lake peregrines in my book, *My Wild Life, A Memoir of Adventures within America's National Parks*, thusly:

> One July day, a pair of peregrines played catch with a prey species of some sort, probably a ground squirrel. The birds took turns dropping and catching their prey, diving, rolling, and sometimes somersaulting all the while. Every few seconds one or both would let out a loud, piercing scream. Such actions were most likely part of their courtship. Those Crater Lake peregrines remain one of my most cherished peregrine memories.

Another of Crater Lake's memorable raptors that I encountered was a great gray owl. Bob Scott and I had driven to Red Cone Spring in the early morning. We had walked around the area for an hour or more before I discovered the owl perched in the open on a heavy limb near the top of a mountain hemlock. It remained still, allowing us both a great view through binoculars. Its huge size, brownish-gray plumage, heavy ringed facial disks, and yellow eyes were obvious. As long as we stayed put, it remained still, even though it was staring directly at us. But as soon as we began to move closer, it flew to another hemlock and perched so that its head was out of sight, I suppose thinking it was hidden from our view. After studying it a little longer, we moved closer, but it then flew away into the adjacent forest. What a marvelous bird!

Mount Shasta, by Betty Wauer

Mount Shasta is located south of Crater Lake, at the southern end of the Cascade Range in north-central California. Shasta is 14,179 feet in elevation, and possesses dual cones. It is what is known to geologists as a "stratovolcano." Although it is considered dormant, hot gasses and fine ash still arise from the summit. The last eruption was about 3,000 years ago.

I have climbed Shasta twice. The climb usually is a three-day affair. The first day involves leave the parking lot off US Highway 5 and packing in to the "climber's hut," staying overnight, and heading up very early the next morning. Leaving early is necessary before the snow begins to soften, making the hike far more difficult. Leaving the hut around 4:30 or 5:00 AM allowed us to reach the summit by about 1:00 to 2:00 PM. The return trip to the hut takes the remainder of the day.

My first climb of Shasta was with Lyman Jones, my Lutheran pastor from Petaluma. I remember very little about details of that climb, with one exception. While returning from the summit, following an ice shut, sitting on my butt and using my ice-axe as a rudder, I hit an extra icy area and lost control. I slide for an estimated 200 yards. I lost both of my clamp-ons and my ice-axe went sliding down the shut another fifty yards or so; it took several minutes before I was back in control.

My second climb was while attending Santa Rosa Junior College, my climbing partners were Dr. Clark Nattkemper, my biology professor at JC, and his friend Eric Spaulding. We spent considerable time at the summit for photographs and just admiring the amazing vistas. An article about our trip, along with a photograph, was later published on the front page of the Santa Rosa Press Democrat newspaper.

Mt. Shasta climb, Clark Nattkemper & Eric Spaulding

Mt. Lassen, by Betty Wauer

To the south is Lassen Volcanic National Park, also within the Cascade Range, and in a way is a sister-park to Crater Lake. But Lassen has not blown its top, although it is very obviously volcanic. This 10,457-foot-high peak is still active; smoke and steam are constantly released. Lassen is one of the few active volcanoes in the greater Pacific "Ring of Fire," a ring of volcanoes that encircles the Pacific Ocean. Lassen last erupted in 1915, when it blew an enormous mushroom cloud some seven miles into the stratosphere.

Today, the scene is of a rather mild forest with a gray volcanic cone sticking out above the greenery. My favorite site in Lassen is Manzanita Lake, a reasonably small lake that is surrounded by chaparral habitat where I found a covey of mountain quail. It is one of the few quail species that undergoes an altitudinal migration. It apparently uses the Manzanita Lake area in spring because it is often one of the areas to be snow-free. It also feeds on ripe manzanita berries in summer. I wrote about them in *Birds of the Northwestern National Parks*:

> Mountain quail are extremely shy and normally run away instead of taking flight when disturbed. Therefore, it is difficult to get a good look at these large quail. Once found, however, they are easy to identify by their long straight black plumes, gray chests, and chestnut throats and flanks, with bold white bars. Often, they are detected first by their loud calls...a rapidly repeated "kup kup kup," and in late summer when a covey of mountain quail is gathered under cover, there issues a medley of clucking, mewing, and whining sounds, mixed with harsh squawks and a "ka-yak, ka-yak" like guinea fowls.

Manzanita Lake, Lassen NP

Still another fascinating bird at Manzanita Lake was the spotted sandpiper. I was first attracted to this bird by its loud "weep" calls. But what is most interesting about this sandpiper is that females arrive on their breeding grounds and establish territories before the males arrive. Once the males arrive, the polyandrous females will actually form their own harem, occasionally competing with other females in fierce combat.

Multiple mates permit the combative female to lay up to five clutches of four eggs, which are then cared for by her male consorts. Apparently, their pioneering method of reproduction has served this robin-sized, spot-breasted sandpiper very well.

Lassen is the southernmost national park in the Cascade Range; it provides an interface with the mighty Sierra Nevada Range of California that contains Yosemite, Kings Canyon, and Sequoia national parks.

Spotted Sandpiper

Chapter 4

The Sierra Nevada and Panamints

The Sierra Nevada Range lies totally in California. It includes three amazing national parks: Yosemite, Kings Canyon, and Sequoia. And a few miles east of Sequoia, across the arid Armargosa Valley, is the Panamint Range.

Yosemite National Park undoubtedly is one of America's most favored parks. Yosemite Valley contains some amazing waterfalls – Yosemite, Bridalvail, Vernal and Nevada – yet two massive granite monoliths seem to receive the greatest attention. Half Dome soars 4,737 feet above the valley floor, and the face of El Capitan is about 3,000 feet from its base to its summit.

Yosemite NP

El Capitan, Yosemite NP

I have had a long-term relationship with Yosemite. All during the years that I lived in Petaluma, Santa Rosa, and San Jose, I made numerous trips to Yosemite. At all times of the year. I recall camping in the Valley one winter night and waking in the morning with three inches of snow on the ground and a half-inch of snow on my sleeping bag. Breakfast that morning was not a pleasant affair. And I also had a number of first-time experiences there. For instance, I wrote about encountering my first-ever dipper in *My Wild Life, A Memoir of Adventures within America's National Parks*:

> On February 28, I discovered an American dipper along the Merced River. I watched it for several minutes as it searched for insects along the icy bank. It actually dove below the icy waters on several occasions, even under the ice-crusted fringes. Through binoculars I could see it literally "flying" below the surface, using strong wing beats to propel itself along the river bottom in its search for food. The American dipper is our only truly aquatic songbird.

I also had an unexpected encounter with a bald eagle. One winter afternoon while driving into the park, where I was attending a training program, I encountered a bald eagle flying toward me. I included that experience *in My Wild Life* as well:

> I suddenly saw a mature bald eagle, bright white head and tail, flying right over the roadway coming in my direction. I was so surprised to see it that I craned my neck to watch it pass overhead. A second later, when my attention was again focused on the road ahead, I discovered a California highway patrol vehicle coming my way; I was halfway over the white line. At the same time, the red lights began flashing on the cruiser. I immediately pulled off the right side of the road, and the patrolman parked across the road in another puff-off.

Bald Eagle, by Greg Lasley

I quickly jumped out of my vehicle, binoculars in-hand, and looked skyward, and yelled at the officer, "Did you see the bald eagle?" He immediately responded, "No, where?" He seemed as excited as I was. He grabbed his own binoculars, and we spent the next ten minutes tying to relocate the bird. We did not, but he apparently believed me. We then spent another fifteen minutes or so introducing ourselves and talked about birding and my participation in the training program. As we parted, almost in passing, he got serious, and said, "Ro, by the way, next time you see a bald eagle over the roadway, be more careful."

The Yosemite training program occurred early in my National Park Service career. It was a three-month program designed to introduce new employees to the National Park Service and to provide training on a whole array of responsibilities. It also provided an opportunity to meet several other new park rangers. I made several friends, and thoroughly enjoyed being in Yosemite Valley.

Each morning, before classes, I walked the valley roadways, once they were plowed enough, and I was able to take some outstanding photographs. Deer were especially abundant in the open meadows, in places where they could find grasses below the snow. But my most exciting find was a great gray owl sitting on a low snag one hundred feet or so off the roadway. Through binoculars, I could see that it had just captured a rodent of some sort, and it was just beginning to tear it apart. But, when it noticed me watching, I apparently spooked it and it flew across the meadow into the adjacent forest. That was my first sighting of a great gray owl, one of nature's most elusive raptor.

South of Yosemite, and perhaps the next most popular park in the Sierra Nevada, is Sequoia National Park. Sequoia contains two extraordinary features: Mt. Whitney, at 14,494 feet, the highest mountain in the contiguous United States, and the General Sherman tree. This amazing sequoia tree is recognized as the largest single-stem living tree on Earth. It stands 275 feet and is over 36 feet in diameter at the base. It is estimated to be around 2,300 to 2,700 years of age.

My experiences in Sequoia are numerous. During my National Park Service career, I was friends with Dick Burns, the park's chief naturalist, and when I visited Dick, he drove me around the park and we hiked the Moro Rock Trail. The trail is a steep quarter-mile staircase of about 350 steps to the summit that offers a spectacular view of the Great Western Divide and the western half of the park.

On another visit to Sequoia, I entered the park from the east side on the Whitney Portal Road from Lone Pine. This is a popular approach to a Mount Whitney Trail that provides access into the highlands. My hike that day, about five miles as I remember, led to a beautiful bright-blue lake surrounded by gray scree slopes. The only wildlife I remember from that day was a number of Clark's nutcrackers. Their "caw" calls echoed across the rather barren slopes.

Mount Whitney is visible from the summit of Telescope Peak (11,049 feet), the high point of the Panamint Mountains that forms the western edge of Death Valley National Park. The lowest elevation in the Valley, the lowest place in the Western Hemisphere at 282 feet below sea-level, is Badwater. The elevation difference between Badwater and Telescope Peak is 11,331 feet, making Telescope Peak the "tallest" peak in the Continental United States. Telescope Peak is the only peak with a base below sea-level.

Lone Pine Lake, Sequoia NP

Panamint Mountains from Dante's View, Death Valley NP

During the six years that I worked in Death Valley, I was especially interested in the birdlife, and so I surveyed numerous areas throughout the park, especially the Panamint Mountains. I was able to hike from one spring to the next that were located in a line at about 3,500 feet along the eastern slope of the Panamints. I eventually published my finding in an article titled "Ecological Distribution of the Birds of the Panamint Mountains," in the ornithological journal *Condor*. My summary statement reads:

> Of 144 species reported for the Panamint Mountains, 75 are considered to be breeding. Three species nest in the valley alluvial fans, 16 species nest in the lower canyons, 11 species nest in the open sage flats and valleys, 41 species nest in the pinyon-juniper woodlands, 19 species nest in the limber pine association, and 8 species nest in the bristlecone pine association. Only one species, the Rock Wren, was found to nest in all of the associations. Forty-three species were found to nest in one association.

But the park contains much more than birds! Bighorn, perhaps receive more attention than any of the other wildlife, although coyotes and kit foxes also get their due. And snakes, especially rattlesnakes, get equal attention. One of the least know is the speckled rattler, or the very red form that is locally known as the Panamint rattlesnake, that occurs only in the Panamint Mountains. The sidewinder, or horned rattlesnake, is found only in the lowlands. I wrote about one of my encounters with this little poisonous snake in *My Wild Life*, thusly:

Rock Wren

The most startling (encounter) was with a sidewinder at the Mesquite Flat Sand Dunes. I was down on all fours looking into a recently excavated badger den when I suddenly detected a flickering tongue of a curled sidewinder within inches of my face. My backward movement was automatic and incredibly swift; I found myself five feet away sprawled out on my back in the sand in a split second.

In summer, I favor the Panamint Mountain highlands. While the Valley can be 110 to 120 degrees on a typical summer day, the mountain highlands rarely exceed the mid-80s. Temperatures normally drop three degrees Fahrenheit for every 1000 feet of elevation. A roadway runs from Mahogany Flat to a series of huge, stone Charcoal Kilns at 8,600 feet elevation, then passes through a big sage flat and pinyon-juniper and mountain mahogany woodlands. Beyond, to the summit is a scattering of limber pine and attractive reddish buckwheat shrubs. From Mahogany Flat, the trail to the summit is 4.2 miles.

Although the Panamint Mountains are hot and dry most of the year, winter storms can create dangerous conditions in the uplands. I recall flying, in a small Cessna, back and forth along the summit trail, searching for a hiker who had been reported lost, without success. I had never before been so buffeted by the thermals that arose from the warmer slopes. The hiker's body was not found until spring.

Telescope Peak in winter

Directly east of the Panamints, only 25 miles from Las Vegas, are the Charleston Mountains, part of the Spring Mountains. The high point is Mount Charleston at 11,916 feet elevation. The highlands are accessible from Highway 165 that follows Lee Canyon to the Bristlecone Trailhead. The trail makes a 6.5-mile loop through an area of ponderosa pines and into a bristlecone pine area. Some bristlecones have been dated at more than 4,000 years old. About 300,000 acres are included within the Spring Mountain Recreation Area, containing a lodge, cabins, and ski slopes.

My single visit into the area was in June 1961, when I walked the 6.5-mile loop trail. My only notes from that day listed a number of birds typical of such Great Basin highlands: hairy woodpecker, violet-green swallow, western wood-pewee, Steller's jay, Clark's nutcracker, and red crossbill.

To the northeast of the Panamints is a parallel mountain range, the White Mountains, which rise to 11,500 feet at the summit of White Peak. Most of the plants and animals that occur there are the same as those in the Panamint and Charleston mountains. However, the White Mountains are famous for a bristlecone pine, named "Methuselah," that is the oldest known living tree in the world, at 4,852 years old.

My visit to see Methuselah was accessible from the town of Big Pine and a grated seven-mile-long rather steep and winding roadway. Methuselah reminded me of many other twisted pine trees found in the highland peaks throughout the Great Basin. Earlier tree-ring samples had dated it, and Methuselah had become famous.

Also Included within this assembly of highland areas is Wheeler Peak, at 13,161 feet, within the Snake Range of Great Basin National Park, Nevada. This park offers a marvelous 12-mile-long scenic drive. My visit to Great Basin National Park occurred when it was known as Lehman Cave National Monument, and when Wheeler Peak still held a glacier that was visible from the scenic drive. The glacier has since disappeared due to climate change that has reduced or eliminated glaciers and snowfields world-wide.

Wheeler Peak, Great Basin NP

Sagebrush Flat below Wheler Peak

The surrounding lowlands is dominated by a big sage habitat, that at first glance may appear rather arid and dry, but that is where I found my first-ever sage-grouse. This is a fairly large bird with a mottled, brown and golden back, a long, pointed tail, and a black belly. During courtship the male is famous for spreading it long, stiff tail feathers, like a peacock, and rapidly inflating and deflating its air sacs resulting in loud, bubbling, popping sounds.

My sage-grouse sighting, however, was not during courtship; I accidentally discovered a small cover while wandering about on the open flat below the park. Two of the birds took flight, and on looking about further, I discovered at least three others.

Several years later, while eating dinner with friends living in Pocatello, Idaho, they served sage-grouse, legally shot in season. I must admit, that was one of the very best wild game meals I have ever eaten.

Chapter 5

Sky Islands

The Southwestern deserts contain a number of isolated mountains that rise above the arid lowlands into forested highlands, a few above treeline. Life zones for these sky islands are easily identified by a pinyon-juniper woodland above the desert, usually above 5,500 feet elevation, and a transition zone dominated by ponderosa pine, and moist canyons and forest above that, the Canadian zone of Douglas-fir and aspen. Sky islands are scattered within the states of Arizona, New Mexico, Utah, Texas, and northern Mexico.

The high isolated plateaus in Zion National Park provide the very best example of sky islands. The Great White Throne is a marvelous example of a sky island. Topped by a luxuriant forest environment, three sides of the Great White Throne are bare of vegetation and seem to highlight that magnificent cathedral. Located along Zion Canyon, it commands all of the park. Zion Canyon and the many other canyons offer a twisting maze of diversity, where cool, protected niches of highland vegetation often lie below arid slopes, dominated by desert vegetation.

Great White Throne, Zion NP

Another of the special features of Zion National Park is The Narrows. This is a deep, narrow canyon in the plateau (generally above 9,000 feet elevation), where the Virgin River, over millions of years, has cut a deep trench into the Navajo sandstone cliffs that rise high overhead. During the years that I worked as a park naturalist at Zion, I waded The Narrows on two occasions.

The Narrows Trail begins outside the park, not far from the park's East Entrance, and follows the bed of the Virgin River for sixteen miles; much of the route requires wading. Few routes compare with this unique hike from the headwaters of the North Fork to the Temple of Sinawava. There is one place where the walls are only twenty-five feet apart and rise more than 2,500 feet to the rim of the plateau. I was so impressed with The Narrows that I wrote an article on one of my Narrows hikes that was published by *Summit Magazine* in March 1965. I wrote:

> The sky was completely blue, as seen only along the Utah-Arizona border…I began to try to follow the winding stream course as far as possible without giving in and begin wading. This I only succeeded in doing for about a mile before the banks became too steep and the rocky streamsides became too few. It was then that I became aware of a strange sensation. It had something to do with this downward plunge with the river, it is best to go forward rather than retreat. It also has something to do with the little sign posted at the Temple of Sinawava: "A sudden flash flood drowned five hikers in the Narrows." … You became amazingly aware of the blue sky, and each shadow along the waterway seems to predict a searching glance upward. Yet even this is forgotten as the first narrow passage is approached.

The Narrows, Zion NP

I was somewhat surprised at the lack of wildlife within the upper Narrows, but once Goose Creek was reached it changed considerably. The quantity of deer tracks was evident enough to indicate that the North Fork and Goose Creek is a heavily used deer-route in and out of Zion Canyon. The common bird was the Dipper. Bruce and I watched it feeding upon insects that it found below the surface of the water, and I followed one to a nest of mud and moss built in a rock crevice not far above the waterline. Wading forward for a closer examination, I became aware of the thin squeaks of young birds from the nest. The adult suddenly appeared and flew down canyon, probably in search for food for her tiny offspring.

There are many more Sky Islands beyond Zion National Park. Arizona has long been known for these special places.

Arizona's principal sky islands include the Santa Catalina and Rincon mountains just northeast and southeast, respectively, of Tucson. The Pinalenos Mountains are located several miles northwest of Tucson; the Chiricahuas are further east, near the New Mexico border. The Santa Ritas are directly south of Tucson and the Huachuca Mountains are further southeast, close to the Mexican border

The Santa Catalinas serve as doorsteps for Tucson. From downtown Tucson, one can drive east on Speedway Boulevard before turning left on Houghton and finally right onto the Catalina Highway, past Sabino Canyon turnoff, and into the higher mountains. The highway switchbacks up to the village of Summerhaven and to the Mt. Lemmon trailhead. The 9,157-foot summit is about four miles beyond. I have spent considerable time within Sabino Canyon that contains a rocky stream that attracts lots of wildlife. Watching feeding painted redstarts and mountain chickadees there will not soon be forgotten

Santa Catalina Mts., by Brent Wauer

The Pinaleno Mountains don't get the same amount of attention from nature lovers as do the Chiricahuas and Santa Ritas, but the summit of Mount Graham (10,720 feet) contains a forest where many of the same wildlife species are possible.

The Chiricahua Mountains, south of the Pinalenos, represent a northern extension of Mexico's mountain provinces. The majority of the range of many of the distinctive flora and fauna of the Chiricahuas lie south of the border within the evergreen Madrean forest and woodlands of the Sierra Madre Occidentalis. Apache and Chihuahua pines, Chihuahua fox squirrel, and mountain (Yarrow's) spiny lizard are Mexican species in the Chiricahuas that barely enter the United States. This area boasts the greatest diversity of flora and fauna of any of the sky islands. *Chiricahua* is said to be an Opata Indian term for "mountains of the wild turkeys."

Pinaleno Mountains, by Brent Wauer

Painted Redstarts

Chiricahua Mountains, by Brent Wauer

The high point of the Chiricahuas is Chiricahua Peak at 9,759 feet, within Coronado National Forest. Access to key sites, such as Cave and Bonita Canyons and Rustler Park, is provided by paved and/or maintained roadways. And access to Massai Point and the trailhead to the summit is provided by an eight-mile scenic drive.

Rustler Park is dominated by a lovely meadow surrounded by a forest of Douglas-fir and ponderosa pine. I spent a cool, enchanting afternoon there, admiring the abundant wildflowers and bird song. One bird that was most memorable was an olive warbler that sang a song that reminded me of tufted titmice. I watched it gleaning insects off the high foliage of a Douglas-fir. And on a couple occasions it flew out to capture a flying insect of some sort. Although I had encountered this warbler elsewhere in the past, this was my first time to be able to admire it for an extended period of time. Its head was almost orange in the sunlight that day.

Mexican Jay

About 12,000 acres of the Chiricahua Mountains are included within Chiricahua National Monument, and 87 percent of that area is designated wilderness. I wrote about one visit to Chiricahua National Monument in *Birding the Southwestern National Parks*:

> One morning in fall, when ripe acorns hung from the oaks after many of the summer resident birds had already departed for warmer climes, I visited Chiricahua's Faraway Ranch. Harvest time was in full swing. The cacophony of Mexican jays, acorn woodpeckers, northern flickers, white-breasted nuthatches, and bridled titmice was audible from as far away as the parking area. As I approached the ranch buildings, I realized that the majority of the birds were centered on the tall oak trees. The drum of the acorns falling on the tin roofs added to the uproar.

The Santa Ritas, located about 40 miles south of Tucson, is a typical sky island that encompasses extensive woodland and forest habitats. The high point is Mount Wrightson that rises to 9,453 feet elevation. Madera Canyon intersects the varied habitats and contains cabins and a small store in the lower portion, and trails provide access into the upper reaches. Because the Santa Ritas are so near Mexico, it attracts a number of Mexican birds not found or are rare elsewhere in the United States. A grand total of 256 species are listed, including 15 species of hummingbirds and such specialty birds as Arizona woodpecker, red-faced warbler, and elegant trogon. Another of the specialty birds is the whiskered screech-owl; I wrote about finding this owl in Madera Canyon in *Raptors, Hawks, Eagles, Kites, Falcons and Owls*:

> My first encounter with this owl was years ago when Bruce Moorhead and I spent a June afternoon and evening wandering along Madera Canyon. It was a couple hours after dark before we heard our first whiskered screech-owl. By walking slowly toward the sound, we arrived beneath an oak tree where the bird was calling. It then took a considerable amount of time, and imitating the bird's song, before we located the owl almost directly above our heads, about a dozen feet away. It seemed unafraid and remained in place looking directly at us for a good two or three minutes before it flew up-canyon and out of view. Although the bird looked very much like the western screech-owl we had seen earlier, we identified it as a whiskered screech-owl primarily due to its Morse-code-like song.

Santa Rita Mountains., by Brent Wauer

Huachuca Mountains, by Brent Wauer

Another of Arizona's sky islands is the Huachuca Mountains, located just south the town of Sierra Vista. The high point in the Huachucas in Miller Peak at 9,585 feet, all within the 20,190-acre Miller Peak Wilderness. There are a total of 50 trails in the area; the Crest Trail, beginning at Montezuma Pass, makes a 9.6-mile loop through mid- to higher elevations. Richard Bailowitz and Hank Bodkin provided a good description of the Huachucas in *Finding Butterflies in Arizona*, thusly:

> The greater Huachuca area – including both east and west flanks, Parker Canyon Lake, and the grass-covered bajadas down to the San Pedro and Babacomari rivers – has a butterfly species list of over 180 species – more than half the state's total! Coronado National Memorial at the south end of the range and the Nature Conservancy's holding in Ramsey Canyon in the heart of the range have both drawn visitors interested in watching and photographing the area's specialties, be they butterflies, birds, or otherwise.

My favorite areas in the Huachucas for both birds and butterflies are Garden and Ramsey canyons. Garden Canyon is entered through the Fort Huachuca Military Reservation; one must check in at the Military Gate before continuing up the canyon. Access roads to Ramsey, as well as Carr, Miller, and Hunter canyons, lie to the south off Highway 92. Ramsey Canyon has been called an "ecological crossroads" where the Rocky Mountains and the Sonoran and Chihuahuan deserts all come together.

The Nature Conservancy operates a site in upper Ramsey Canyon that has become one of the best places in the U.S. to see hummingbirds. The Conservancy also provides guides that help visitors enjoy the area and its wildlife. My visit to Ramsey Canyon, where I hiked up the trail a couple miles above the Conservancy headquarters produced a number of specialty birds, including Arizona woodpecker, elegant trogon, hepatic tanager, painted redstart, and red-faced warbler.

Red-faced Warbler

The red-faced warbler was especially appealing with its red face, black crest with a narrow white collar, gray back, and contrasting white belly. I watched it feeding in the outer branches of the conifers, with constant small jerks of its tail. This neotropical warbler spends it winter months in western Mexico and Central America, but returns to its nesting grounds by early April. Then, unlike most warblers, it constructs its nest of pine needles, fine bark, and soft plant materials in depressions on the ground, concealed in grasses or sheltered by rocks or logs. And, I was a bit surprised at its clear and penetrating song, whistle notes that can be described as "a tink a tink, tsee tee, tsee, tswee, twsee."

Carr Canyon may be considered the heart of the Huachucas; there are two campsites along the roadway that runs up into a pine forest and a trailhead at about 7,000 feet elevation. This is the best and easiest access to the Huachuca high country.

In southern New Mexico and southwestern Texas, three mountains rise out of the Chihuahuan Desert lowlands into the highlands. North to south, they include the Guadalupe, Davis, and Chisos mountains.

Guadalupe Mountains, the northernmost area, are formed by a long north-south range of Permian limestone in southeastern New Mexico and adjacent Texas. The northern Guadalupes are lower in elevation and drier than the southern half of the range which forms a magnificent escarpment at the southern end. From north to south, the range ascends gradually to 8,749 feet at the summit of Guadalupe Peak, which is the highest point in Texas. The eastern slope is heavily dissected by numerous canyons, while the western escarpment is very steep with open alluvial fans. Approximately 65 square miles lie above 5,500 feet elevation.

Guadalupe Peak, Guadalupe Mountains NP

One July day, I hiked the very steep Bear Canyon Trail to The Bowl, a grassy meadow just below the summit, where I camped for two nights. The Bowl, about 100 acres in size, contains a relict forest representing cooler times. I found several mountain birds there that rarely occur at lower elevation: whip-poor-will, western wood-pewee, Steller's jay, mountain chickadee, house wren, warbling vireo, Grace's warbler, and dark-eyed junco.

The most obvious of these was the Steller's jay, an all-blue jay with a blackish-blue crest. A pair of these birds spent considerable time watching me as I prepared my camp, and their loud, grating calls were evident throughout my stay. Spotted owls, listed as endangered, also occur in the Guadalupes.

The more popular area in the Guadalupes is McKittrick Canyon that follows a stream that is shaded by bigtooth maples and cottonwoods. One early spring, I found the area alive with birds. Although the lower slopes were dominated by desert vegetation and desert birds, the scene changed dramatically after about one mile. Black-chinned and rufous-crowned sparrows, canyon towhees, and ash-throated flycatchers were numerous. And I was somewhat surprised to find a pair of magnificent hummingbirds. After a couple additional hikes up McKittrick Canyon at a later date, I realized that this habitat was perfect for this large, colorful tropical hummer. Nowhere else in the U.S. is it so common.

McKittrick Canyon

McKittrick Canyon also is where elk were reintroduced to the Guadalupe Mountains in 1928. Elk were fairly common in the early 1800s, but had been hunted out by the late 1800s. Local rancher J.C. Hunter captured a breeding stock from the Black Hills of South Dakota and released 44 individuals into McKittrick Canyon. At least a few individuals spread into the highlands, including The Bowl, within a few years. Today, according to National Park Service records, the population is estimated to be between 30 and 40 animals.

The Davis Mountains lie only 100 miles southeast of the Guadalupes. They are of igneous origin, but form rolling hills and open, grassy valleys at mid-elevation, and rocky canyons and jagged peaks in the highlands. Approximately 70 square miles lie above 5,500 feet. The high peak is Mount Livermore at 8,353 feet elevation; nearby Pine Peak is 6,800 feet. On one climb on Pine Peak, with friend Greg Lasley, we discovered an Ursine giant-skipper perched at the very summit; Greg took several photographs of that rare butterfly.

And to the east is Mount Locke (6,790 ft.) with the McDonald Observatory, accessible off Highway 118. When my friend Mark Adams managed the site, he and I spend several days together searching for birds and butterflies.

Mt. Livermore, Davis Mountaints

During the years that I worked at Big Bend, I visited the Davis Mountain on numerous occasions, including participating in several Davis Mountain Christmas Bird Counts that covered a goodly portion of the highlands. On December 27, 1970, I recorded a total of 55 species. Kelly Bryan was the coordinator of that count which included a huge area within and around the Davis Mountains.

I also attempted to capture enough Montezuma quail, fairly common on the grassy slopes, to reintroduce them into Big Bend. I was unable to capture a sufficient number there, but in January 1979, with the help of David Brown and his dog, I was able to procure sufficient numbers in Arizona's upper grasslands. I released them in Pine Canyon where they spread throughout the Chisos highlands. Recent sightings have been reported for the South Rim.

Years later, I guided nature trips into the Davis Mountains. And when I became a butterfly enthusiast, I spent considerable amount of time searching the Davis Mountains for butterflies. The Lawrence E. Wood Picnic Area and surroundings was a superb place to find a goodly number of butterfly species.

Still further south, just south of Fort Davis, is the Chihuahuan Desert Research Institute. The grounds contain a headquarters building with a gift shop, a series of green houses that are managed by botanists at Sul Ross State University in nearby Alpine, and self-guided nature trails. Walking the Modesto Canyon Trail, with its narrow stream, offers several fascinating floral surprises.

Montezuma Quail, male

Davis Mts.; notice McDonald Observatory on two high points

Emory Peak, Big Bend NP

The Chisos Mountains, which form the core of Big Bend National Park, lie about 100 miles southeast of the Davis Mountains, and are the southernmost mountains within the United States. The Chisos are an igneous mass of intrusive and extrusive rocks that rise out of the desert lowlands to 7,835 feet elevation at the summit of Emory Peak. The Chisos peaks, canyons, foothills, and alluvial fans are more rugged than the Davis Mountains, but less than 10 square miles lie above 5,500 feet elevation.

A few high, moist canyons in the Chisos Mountains contain secret places with mountain maples, the same big-tooth maples that produce magnificent fall color displays in the Rockies and Sierras. Texas Madrone, Arizona cypress, Arizona yellow (or ponderosa) pine, and endemic Emory and Grave's oaks share these high canyons. The adjacent slopes produce a colorful groundcover of scarlet bouvardia, sleepy catchfly and cardinal flowers. Black bear, Carmen Mountain white-tailed deer, and mountain lions find cool ponds of water in their depths, and overhead Colima warblers, Mexican jays, hepatic tanagers, and blue-throated hummingbirds forage for sustenance. And high overhead can be found such avian species as peregrine falcons, zone-tailed hawks, and white-throated swifts.

Black bear are surprisingly common today in the high Chisos, but when I moved there in 1966, bears were totally absent; they had been shot out during the ranching years before national park designation. About 1968 or 1969 they began to return, undoubtedly because the bear population in the neaby Maderas del Carmen was not as threatened as earlier and was increasing.

Bear at Big Bend

I had a number of bear encounters during my tenure at Big Bend. The photo above was taken one day at the Sam Nail Ranch, where I had stopped to check the birdlife. A windmill pumped enough water to form a puddle under the vegetation, that included a tall pecan tree. I entered the grove and sat on a bench with a vew of the puddle. Suddenly a bear, which had been lying in the puddle, jumped out and ran to the tree, passing me only about 10 feet away, and climbed part way up the tree. I was able to photograph the bear from about 50 feet as it was clinging on the tree trunk.

Another memorable bear encounter, with an adult and her cub, occurred while I was leading a seminar group in Big Bend's Boot Canyon. I included that experience in *Wild Critters I Have Known*:

> They were moving up-canyon toward our group, en route no doubt to a waterhole that we had just pased. I immediately stopped to let the sow realize our mutual predicament. We were on her path to water, and she was on our route back to Panther Pass and the Chisos Basin. A sight breeze was blowing toward us; she at first seemed unaware of our presence. I puposefully shuffled my feet in the loose rocks to alert her to our position. She stopped then, staring our way, and sniffed the air to detect the source of the noise. It was soon obvious that she was aware that her route to the water was blocked and that one of us had to give in.

Bear in trail, by Jim Weber

She first made several strides toward us, expecting, I think, that we would retreat. I slowly advanced toward her, making considerble noise with my feet, and even picked up a large flat rock and letting it drop. It was then that she realized, apparently, that she might need to reconsider her attempt to get her way. But she continued to bluff her way toward us, even growling as she swung her body back and forth. Through binoculars, I could see her beady eyes focused on us.

I continued to move, step by step, ever so slowly toward her, making shuffling noises in the rocks. Finally, after about ten minutes, she began a slow retreat up the slope, following her cub. But we watched her carefully for any movement that she might make toward us. By then, I am sure that she realized that we did not pose a danger but only wished a safe passage down-canyon. She remained on the hill, sitting on her haunches, watching us pass and disappear from view.

The Big Bend Country has experienced a history of natural change. Phillip V. Wells found that he could identify and date the area's past and present vegetation through the study of wood rat deposits found in caves and other dry places. Excavations of numerous *Neotoma* "middens" from the foothills of the Chisos Mountains and Burro Mesa, down to the surrounding lowlands, resulted in some startling discoveries. Radiocarbon dating revealed that 15,000 to 20,000 years ago the vegetation of the desert valleys, such as those places along Marvillas and Tornillo creeks, was similar to that of the Chisos Basin today.

Wells theorized that conditions during the Pleistocene were considerably moister than today. Approximately 10,000 years ago the climate began to change from one of relatively high precipitation to one of low rainfall. One can therefore assume that the Chihuahuan Desert of West Texas is rather recent (in geologic time), having developed only during the last 10,000 years.

Giant Dagger, Yucca faxoniana

Climate change may largely explain why the Chisos highlands still holds a number of Mexican plant and animal species that are rare or not present further north. Examples include false agave or hechtia, candelilla, giant dagger, in the lowlands, and Havard agave, Mexican pinyon, drooping juniper, and several oaks in the highlands.

The Carmen Mountains white-tail deer is another example. Prior to the time when the climate included greater rainfall, Carmen white-tails were the common deer species in the Greater Big Bend Region, but as drier conditions increased, these deer deserted the desert lowlands and moved into the cooler highlands. About 10,000 years ago, mule deer began to move into the lowlands, leaving the white-tails isolated in the Chisos Mountains, the Maderas del Carmen, and a few other highland areas along the border.

An additional relict is the Colima warbler, a neotropical songbird that nests in the Big Bend, but nowhere else in the U.S. It also is a nester in the Maderas del Carmens and a few scattered highlands to the south. It migrates southwest to the Mexican states of Jalisco and Colima for the winter. I wrote about my first sighting of this bird in *Birding the Southwestern National Parks*:

> I could hear the song ahead of me, to the right of the trail. There was silence for a few minutes, broken only by the almost ubiquitous calls of Bewick's wrens and the hoarser notes of Mexican jays. Then I heard the song again, closer now and coming from the little canyon along the trail. I stopped and waited for the little yellow and gray bird that I expected would soon move into the higher branches of an oak that grew taller than the surrounding brush. Suddenly there it was: my first Colima Warbler! I watched it work its way up and around the Emory oak, gleaning the branches and leaves for insects. It captured a long green caterpillar and for just a second or two seemed to examine its breakfast before swallowing the prize. The Colima Warbler fed there in the sunlight for several minutes, and every thirty second or so it would put its head back and sing, a song a little like that of a Yellow Warbler but shorter, faster and less melodic.

Colima Warbler, by Greg Lasley

All during the six years that I worked in Big Bend National Park, I led trips into the Chisos to show participants this unique warbler. Also, during those years I spent considerable time studying and censusing Colimas. In *A Field Guide to Birds of the Big Bend*, I wrote about the species thusly:

> Colima Warblers usually arrive in BBNP during mid-March…In dry years they are extremely difficult to find until the first or second week of April. Afterward, there is little problem in finding birds in the proper habitat. They normally are vociferous; males sing throughout the days prior to nesting but only in the early morning hours and on a few other occasions during the day when nesting. Nests are placed on the ground in leaf litter or under clumps of grass. Both adults build the nest, incubate the eggs, and care for the young. I have banded nestlings ready to leave the nest from May 25 to July 6. Very little attention is given the youngsters after they have been out of the nest a few days.
>
> Although there is a noticeable decline after mid-July, a few Colimas usually can be found in choice sites through August and early September; my latest sighting is September 18.

Chapter 6

Mexican Highlands

Forty miles southeast of Big Bend's Chisos Mountains, the Sierra del Carmen form an impressive mountain system containing several peaks over 8,000 feet elevation. The Sierra del Carmen is the northern-most portion of the Sierra Madre Oriental that runs almost to the Isthmus of Tehuantepec. The northern half of the del Carmens consists of tilted limestone layers, mostly within the United States. The southern portion, known as the Maderas del Carmen, is located in Mexico and is of igneous origin. The result is a very different landscape.

The western escarpment of the Maderas is steeply faulted and forms magnificent cliffs and deep canyons. The highest of these peaks is Loomis Peak at 8,960 feet elevation. The eastern slope is gradual and contains numerous rather gentle but dissected canyons. Approximately 115 square miles lie above 5,500 feet elevation.

Maderas del Carmens from Maderas Peak

While working in Big Bend National Park, I was able to visit the Maderas del Carmen highlands on several occasions. A number of old roads, once used to drag lumber to dusty roadways, provided superb hiking trails. The view from the summit of Loomis Peak may be one of the finest in North America; a great place for photographs! The desert lies to the west, below the 5,500-foot escarpment. On a clear day the Chisos Mountains stand out to the northwest; they seem almost touchable although more than fifty miles away. On dusty days they glimmer like a mirage against the far horizon. To the north are the other peaks of the Maderas del Carmen, and the flatland that lies beyond is the uplifted limestone portion of the range.

I wrote about one truly memorable incident that happened atop Loomis Peak, in *Birder's Mexico*, thusly:

> We heard the peregrine tiercel (male) calling to his mate long before we saw him. We also heard the responsive cries of the peregrine haggard (female) somewhere on the face of the gigantic cliff directly across from our lofty perch. I turned my binoculars toward the deep canyon below and searched for an invisible dot that would be the eyrie-bound hunter. Another call from the depths below somehow directed my search a little to the left. There he was! The tiercel was flying swiftly in a direct course towards the eyrie and his waiting mate.
>
> I watched that incredible bird with all of the respect he so well deserved. He was returning to his eyrie with food that he had caught in the desert far below. His flight upward was at an angle of at least forty-five degrees. Yet the weight of a bird almost his own size – either a white-winged or mourning dove, I couldn't be certain – did not seem to hamper him. The peregrine's powerful wing strokes drove him upward.
>
> Three of us watched the tiercel every stroke of the way after my first discovery. We knew when he was nearing the eyrie because of the increased calling of both birds. Then suddenly he disappeared from view into a hidden crevice within the rhyolitic fortress. We understood what was taking place as he shared his kill with his mate and perhaps, three to four downy youngsters.
>
> A few minutes later the tiercel reappeared flying out and upward toward the top of the cliff another hundred feet or so. There, in full view and perched on an old weathered snag, he rested. We watched him preen, ruffle his feathers, and finally settle down as if to stand guard on his Maderas del Carmen homeland.

Peregrine Falccon

Another trip into the del Carmens highlands occurred in April 1972 when I led two scientists, from the University of Arizona, into the Carmens to take tree-ring samples to date the forest. Chuck Stockton and Marvin Stokes stayed at my home in the park the first night, and we hired Senior Padillo of Boquillas to drive us to Los Cohos Spring the next day. We camped there for three nights before Senior Padilla returned and hauled us back to Boquillas. We wandered all around the highlands during those three days.

Tree Ring sampling in Carmens by Chuck Stockton & Marv Stokes

My most fascinating discovery on that trip was finding three ponderosa pines with large, oval holes that had been excavated by a woodpecker. One of the holes was reasonably fresh; it was larger than any excavation from a hairy woodpecker, the only known woodpecker in the Maderas. The one possibility in my mind was a pileolated woodpecker. However, after returned home, and discovering that pileolated woodpeckers do not occur in northern Mexico, I thought that it was more likely that of an ivory-billed woodpecker. Another Carmen trip was essential. I included details of the follow-up Carmens trip in *Birder's Mexico*:

> Three weeks later I returned to the Maderas del Carmen highlands to try to find the bird or hard evidence of its existence. I found additional nesting trees, although every one was old and not adequate proof. I attempted to climb to one of the nest holes, hoping that an ancient feather had been left behind. But a near fall reduced my enthusiasm for that method of discovery.
>
> One more piece of circumstantial evidence came my way on that second trip. I met a bear hunter wandering along Madera Canyon early one morning. We struck up a conversation and he soon was telling me about the local wildlife. I learned about the bears and *panteras*, as well as the deer and fox. I opened my Peterson field guide to the plates on hawks and falcons and asked him if any one of those birds lived in these mountains. With only a moment's hesitation he pointed to the Cooper's hawk, goshawk, and peregrine falcon. Although I was a little uncertain of the goshawk then, I later found it nesting in Madera Canyon. I next turned to the plate of western warblers. Again, he was correct. He pointed only at the painted redstart and Colima and olive warblers.
>
> The only plate I had found of the imperial woodpecker at that time was in Ernest Edward's 1968 bird-finding book. I had photo-copied that plate and it had reproduced quite well. I took that plate from the back of my book, unfolded it, and asked the bear hunter if any of the birds on that plate lived in the Maderas del Carmen. He looked at all of the illustrations. Then he pointed at only one, the imperial woodpecker.
>
> I asked him how recently he had seen one. He said it had been a long time, maybe four or five years ago; he said that he used to shoot them for food because they made a very good dinner.
>
> I have returned to the Maderas del Carmen seven times to search for the imperial woodpecker, but since 1970 have found no evidence. My last trip to the Maderas del Carmen was in 1976 and with the same amount of anticipation as I had experienced on that second trip. Two friends – Joan Fryzell and Grainger Hunt – had told me of seeing a "large crested woodpecker" in Canon de Oso that previous year. Joan, Grainger, and I revisited the site in 1976, but I again returned without proof.

Author in Oso Creek, by Betty Wauer

To the south of the del Carmens are an abundance of peaks and deep valleys of the Sierra Madre Oriental, where some of Mexico's unique wildlife occur. The highlands near Monterrey provide one example; my friend Greg Lasley and I visited the area to find the endemic maroon-fronted parrot. They occur only within the northern-most highlands, in the tristate area of Nuevo Leon, western Tamaulipas and Coahuila. I wrote about that experience in *Birder's Mexico*:

> Their loud raucous calls were almost overwhelming. We reached a place at the base of the high cliff where we could sit and watch. Fourteen maroon-fronted parrots were perched only 125 to 150 feet away from us. But in less than a half-hour that number increased to twenty-four, and soon after that there were thirty-eight individuals, more than three dozen very loud and highly vocal birds.
>
> They appeared as curious about us as we were about them. It seemed to be that the newcomers had joined the first group to assess our presence. They openly gawked at us, shuffle a bit, and preened one another as we watched. They were perfectly safe even at only 125 feet away; their perch was on two scraggly Arizona cypress trees growing out of a deep crack at the base of the eighteen-hundred-foot cliff face.

The majority of Mexico's landscape is mountainous. There is an often-told story about a sixteenth-century description when a Spanish king asked a recently returned conquistador for his description of "the new land." He seized a piece of paper, crushed it into a crinkled ball, and then laid it in front of the king. "There, your Majesty," he said, "is a map of your New Spain."

Mt. Orizaba, by Betty Wauer

Five of Mexico's mountains are higher than the highest point in the continental United States, Mount Whitney (14,494 feet). Those five Mexican peaks are part of the magnificent volcanic range known as the Sierra Volcánica Transversal that cuts across central Mexico from Sinaloa to central Veracruz. The highest is Pico de Orizaba (18,851 feet), or *Citlaltepetl*, a Nahua Indian word meaning "Mountain of the Star." Orizaba is located approximately 140 miles east of Mexico City in the state of Pueblo. It is appropriately named as it provides an outstanding landmark that can be visible for more than a hundred miles in all directions.

In January, 1981, after spending several days at Catemaco, my companions and I drove north to explore Orizaba to search for two rare bird species – the endemic black-polled yellowthroat (formerly known as Orizaba yellowthroat) and slaty finch – that Bill Shaldach, a friend living in Catemaco, had once collected on the upper slopes. I included our search in *Birder's Mexico*:

> We followed a ridge for several miles through some very patchy landscape. Abundant and extensive areas of Montezuma pine, smaller patches of jelcote pine with its long drooping needles, cultivated fields of various sized fenced with rockpiles and huge century plants, and low scrubby areas were intermixed...The habitat looked like chaparral areas in the southwestern U.S. to me...In spite of considerable effort spent searching, we did not find either bird.
>
> I climbed to the summit of an isolated hill, situated just below the forest that formed a line of vegetation below timberline, to get a good view of the surrounding landscape. It was well worth the time and energy. The view alone, toward the snow-covered peak that formed a magnificent backdrop to the forest and fields, made the entire day worthwhile.

Central Highlands

Eighty to ninety miles west of Orizaba are the twin volcanic cones of Popocatépetl and Iztaccíhuatl, also part of the Sierra Volcánica Transversal. Popo, at 17,891 feet elevation, has long been one of my most favorite peaks in all of Mexico. I used one of my photos of Popo on the cover of *Naturalist's Mexico*; in the second edition the title was changed, without my knowledge, to *Birder's Mexico*. I provided a brief description of these two peaks:

> The names of the twin volcanoes – Popocatepetl (pronounced "popo-ka-tep-etle") and Izataccihuatl (pronounced 'ic-ta-see-wattle") – can hardly be spoken in the same breath. But both are extraordinary in their aesthetic appeal and the resources they possess. Both peaks have been incorporated into one comprehensive national park. The area contains extensive coniferous forests and grasslands at the base of the peaks, and a boreal forest that extends up to the upper slopes. The highest slopes above timberline are principally scree-covered, while the tops of the peaks are usually under snow and ice in winter as well as during the rainy season from June to October.

A paved roadway goes over a pass, though a dense forest, and ending at Tlamacas, at 12,800 feet above sea level, the park's headquarters. Tlamacas contains a small lodge and dining room for overnight stay, as well as a trailhead to the higher slopes. The pass, located between Popo and Ixta, is known as Paso de Cortez; it is the route where Hernan Cortez and his Spanish army crossed these mountains in November 1519, in route to Tenochtitlan, the Aztec capital, now Mexico City.

We stayed overnight at Tlamacas, registered the following day, and hiked for about four miles to where the trail was closed due to drifting snow. But what a glorious day it was, with amazing views in all directions. And I recorded a number of birds, many of which could just as well have been in the highlands of the Rocky Mountains. Five of those species, however, are found only in Mexico's highlands: pine flycatcher, gray-barred wren, russet nightingale-thrush, fan-tailed warbler, and striped sparrow. The fan-tailed warbler was most impressive; I watched it search for insects along the trail, fanning its tail to startle insects just like mockingbirds and redstarts do in the U.S.

Sierra de Oaxaca

Further south is the Sierra Madre Oaxaca. This high, wild country seldom receives the respect as do the Sierra Madre Occidental or the Sierra Madre Oriental. But the Sierra Madre Oaxaca rises to 11,141 feet at the summit of Cerro Zempoaltepec. I got an excellent perspective of the Oaxaca highlands while driving Highway 175 that runs from Oaxaca city across the Sierras to intersect the Gulf Coast Highway 175. The first 100 or 150 miles north of Oaxaca city is one of the most spectacular routes I have ever seen. I described that area in *Birder's Mexico*, thusly:

> It immediately climbs very steeply up a deep canyon to an open forested plateau, then drops off again into a deep valley, and up again toward the next ridge, and on and on. The habitat in the first several valleys was arid, but good riparian habitat occurred in the bottoms. Further along, as my route continued its gradual climb upward, the valleys provided a more luxuriant growth of tropical deciduous forest. The higher ridges and plateaus that I could see were covered with coniferous forest. Several of the narrow, upland side-canyons provided a barranca-like environment with dense, broadleaf vegetation. Above all of these, at the very top of the great dissected plateau, was a rather arid, mixed pine-oak forest.

I stopped along the roadside whenever possible and walked into the adjacent forest and into a few drainages to see what birds might be present. Several were new Mexican species for me. I encountered a small flock of the endemic dwarf jays at one stop, and a happy wren at another, and perhaps the bird of the day was a male cinnamon-bellied flower-creeper. What a neat little bird it was! A tiny black and deep cinnamon bird with an upturned, hooked bill – the only small Mexican species with such a characteristic. I watched it "snip" its way into the base of a small flower from which it extracted nectar.

At another stop I walked down a deep draw for several hundred yards, where I found a fruiting tree I identified as a hackberry that was full of feeding birds. I sat at a comfortable distant away and watched what took place. Within less than an hour I had added a couple dozen birds to my trip list. Mexican species seen there included spot-crowned woodcreeper, tufted flycatcher, several rufus-backed robins, a pair of Aztec thrushes, ruddy-capped nightingale-thrushes, crescent chested and red warblers, chestnut-collared brush-finch, and a collared towhee.

The Sierra Madre Occidental, or "Mother Mountain Range of the West," extends over 800 miles along the western side of the Mexican mainland. The range is approximately 130 miles in width and averages over 6,000 feet above sea level. A few of the higher peaks exceed 10,000 feet elevation. The mountains are composed of volcanic materials that are often exposed in the deep barrancas, a few of which may exceed 6,000 feet in depth.

Volcan de Fuego

Volcán Colima and its' sister-cone, Volcán de Fuego, rise to 12,500 feet elevation. Dick Russell and I drove above Atenquique, Colima, on a very dusty and winding logging road into the highlands, slowly ascending onto the southern flank of this active volcano. We were amazed at the diversity of habitats above the cultivated fields and arid woodlands. The first change was a gradual one of increased tree size and density that produced a good example of tropical deciduous forest habitat. But that zone was little more than a narrow band, because shortly, at about 5,000 feet elevation, we entered a pine forest that looked very similar to the ponderosa pine forests of the southwestern United States.

At about 6,000 feet the pine forest began to change to an oak-dominated habitat. We encountered a cobblestone road about eight miles above Atenquique that led to a microwave clearing, where we stopped and birded for an hour or so. That is where we found one of our most-wanted birds: a dwarf vireo. I wrote about that encounter in *Birder's Mexico*:

> We had worked our way along the edge of the upper clearing to where we had a good view of the oak woodland. I imitated a ferruginous pygmy-owl call. Immediately a tiny bird flew out of the adjacent oak canopy into an isolated oak within the clearing. It didn't stay long, but sang a very short song that reminded me of a Hutton's vireo song. My first glance at the bird confirmed it as a vireo, but it was smaller than a Hutton's vireo and rather dull with two thin obvious wingbars. Then it was gone, as fast as it had appeared. Dick and I looked at each other for just a second, then both exclaimed together: "That was a dwarf vireo!" We had found one of Mexico's true rarities. That sighting of such a wanted species was what Mexican birding is all about. The joy of finding a rarity in such out-of-the-way localities makes life most worthwhile.

Sierra Madre Chiapas

El Triunfo, located in a wild and remote section of the Sierra Madre de Chiapas, can be reached only with considerable effort. I joined one of the Victor Emanuel's VENT tours that provided the necessary guidance and support. Twelve of us had flown from Mexico City to Tapachula, a small but busy city in the extreme southwestern corner of Mexico. From there we had boarded a Volkswagen microbus and drove north on Highway 200 toward Tonalá. At Mapastepec, 70 miles north of Tapachula, we turned off the paved highway onto a gravel road that ran east toward the mountains.

We continued another eight miles to the end of the road where we met four local cowboys and stock that would haul our gear up the 25-mile trail to El Triunfo. I described that route in *Birder's Mexico*, thusly:

> The first afternoon's hike was confined to the lowlands where temperatures were in the nineties. The trail gradually ascended the western slopes of the Sierra de Chiapas, and temperatures dropped with elevation. On that first afternoon, the four-mile trek along the Rio Novillero, in and out of the tropical deciduous and tropical evergreen forest habitats, taxed our strength and endurance. We all were glad to reach Paval, our camping destination for the night…We covered considerably more ground the next day, our first full day on the trail…Some twelve miles above Paval we reached our campsite at Canada Honda. It was there, just as we approached our camp, that we discovered a small flock of azure-rumped tanagers; one of our most wanted of the many birds we had hoped to find.

Although the azure-rumped tanager is sometimes called Cabanis's tanager, azure-rumped is a much more descriptive name for this lovely species. Even that name, however, does not do it justice. Its head, back and wings are a deep greenish-blue in good light. Its wings are blackish but show bluish color along the edge of each feather. The front of the bird is blue-gray on the throat, gradually fading to much lighter color, almost white on the belly and crissum. And careful study revealed distinct black spots on its chest.

On our third day at about noon, after passing through an evergreen forest, we crossed a pass and began a gradual descent through a mesic forest into an obvious valley containing a forest of huge trees loaded with an amazing assortment of bromeliads and other epiphytes, a cloud forest.. An hour later we walked out into an open area. We had arrived at El Triunfo, the Shangri-la of my dreams. And by late afternoon, after pitching my tent, I was ready for a meal of freeze-dried food. I was greatly anticipating the coming dawn, which I described in *Birder's Mexico*:

> What a dawn it was! There was not a cloud in the sky. I watched the sunlight begin to creep down the hillside, enthralled that I had really arrived at the El Triunfo Shangri-la of my dreams. As I stood there pondering that reality, a male resplendent quetzal in full breeding plumage suddenly shot out from a treetop not more than a hundred yards away. His appearance provided several of us with an unbelievable view of his dazzling green, red and white in a display flight that I had previously only read about.
>
> The tiny village of El Triunfo was located at 6,700 feet elevation and surrounded by higher ridges, some of which exceeded 9,000 feet. The clearing provided us with several Mexican birds of interest. A number of gray silky-flycatchers cavorted about several of the surrounding treetops. The shy blue-and-white mockingbird was mostly evident by songs. A pair of flame-colored tanagers were feeding at bromeliads. Yellow grosbeaks were most numerous and their clear robin-like songs dominated the morning sounds.

El Triunfo's cloud forest is one of the largest and most spectacular remaining in Mexico. Cloud forest habitat is usually found between 5,000 and 7,500 feet elevation. It is maintained by moisture-laden air currents that rise up from the humid lowlands to form clouds and rain in the uplands. The number of rare and endemic flora and fauna is exceptional. Approximately 50,000 acres of the El Triunfo cloud forest was established as a Santuario de Fauna y Flora del Gobiermo in May 1972.

I had developed a list of all the possible birds to be found at El Triunfo, and by the end of our five-day stay I had seen the majority. Perhaps the horned guan required the greatest effort by climbing the high ridge and walking the trail to the summit. I still may not have found it without the local guide who had to point it out to me in the high foliage. I described it in *Birder's Mexico*:

> Horned guans are larger than turkeys. They possess glossy-black plumage on its back, head, tail and belly, but a finely streaked white breast, and a red throat patch that can easily be missed. But the most unusual character is the surprisingly large vermillion-colored spike or horn, or casque that occurs on the crown, just above the eyes. This feature seems to dominate the head so much that the bright yellow bill goes unnoticed.

My stay at El Triunfo was a great success! Specialty birds of the El Triunfo Preserve included two guans, the horned and the highland guan; the white-faced quail-dove, barred parakeets; blue-tailed, wine-throated, and bumblebee hummingbirds, tawny-throated leaftosser, black-throated jays, and several more species of special interest. Plus, a number of wintering neotropical passerines were also present. It was a smorgasbord of colorful birds.

I left El Triunfo with great appreciation and a good deal of emotion. I wrote the following in *Birder's Mexico*:

> It had been a fascinating and exciting several days in probably the most remote place I have ever visited. Exploring its canyons and slopes provided me with more than just another wilderness experience. I was left with a personal commitment to trying to do something more to help protect its fragile ecosystem from further exploitation…I can only hope that El Triunfo is able to live up to its own name, "The Triumph." Being there was indeed a triumphant experience for me and my friends, and led us to hope that Mexico will continue to keep El Triunfo intact for all of mankind. There are no substitutes left.

Index

D

E

F

G

H

I

J

K

L

M

N

O

P

Q

R

S

T

U

V

W

Y

Z